AF540751

ENVIRONMENTAL GEOMORPHOLOGY

By

Dr. Govind Prasad

M.A., Ph.D., D.litt., F.A.MSE (France)

Reader

Dept. of Geography

Rana Pratap Postgraduate College

Sultanpur (U.P.)

(India)

DISCOVERY PUBLISHING HOUSE PVT. LTD.

NEW DELHI-110 002

First Published-2008

ISBN 978-81-8356-265-2

Published by:

DISCOVERY PUBLISHING HOUSE PVT. LTD.

4831/24, Ansari Road, Prahlad Street,
Darya Ganj, New Delhi-110 002 (India)
Phone: 23279245 • Fax: 91-11-23253475
E-mail: dphtemp@indiatimes.com

Printed at:
Arora Enterprises
Laxmi Nagar, Delhi–110 092

new construction. Thus, man's faulty assumptions and mismanaged efforts have germinated the hazard of eco-degradation. The environmental disaster creates the challenge which in turn advances the scientific researches basically on eco-protection in spite of earth description. Now, the worldwide researchers, attention is tossing upon the awareness of environmental quality and the system of geomorphological study and the earth science dialogues have coincided to environmental study.

Eco-degradation is the major problem. A large scale attention is needed through planning processes. The protection of environmental diversity is too much necessary for the integrity of our natural resources. Scientific management of the environment has assumed increasing importance in recent years, all over the world. There is a great awareness now to appreciate and utilise the natural environmental potential for achieving optimum results not only without degradation of what has been gifted by nature but also without affecting the ecological balance. Thus, the theme of the present day researches has oriented towards eco-development where the concept of development is interlinked with conservation and improvement of natural resources. Moreover, of all the environmental sciences, the application of the earth sciences is now considered fundamental to practically any landuse that man can think of. In the light of the above principles, the theme of the present dissertation is derived. The present book, entitled "Environmental Geomorphology is based on the geomorphic unit of Chitrakut and its Adjoining Region" which is a part and parcel of Banda (U.P.) and Satna (M.P.) districts. The region is prosperous in natural beauty and is a well recognised spot in Indian culture due to its association with the Ramayan and Goswami Tulsidas, the famous epic writer. It is famous as the holy centre of 'Lord Rama', popularly known as 'Chitrakut Dham'.

The natural landscape of the region is greatly responsible for its distinctive cultural milieu. The Vindhyan Scarps with numerous springs, ravine tracts, the rivers forming the rapids and waterfalls and the richness of natural vegetation provide the area an excellent scenic beauty which has great potentials for attracting the tourists from all over the world. The great epic 'Ramayana' testifies clearly

Preface

Geomorphology is the study of science of landforms. Landforms are the product of the processes operating with the march of time. Following the classical work of Davis' landform is a function of structure, process and stage. Pioneer researches validate that the geologic structure plays dominant role in the evolution of landforms as a chief controlling factor. One can examine a classic scenery of different types of relief features and a multidimensional landform assemblage over earth surface, because geomorphic processes operate at different styles and rates. The processes also leave their imprint upon landforms wherein each geomorphic process develops its own characteristics assemblage of landforms which must have ornamented in an orderly sequence. The outlook of landscape is the product of a group of processes. Thus, the geomorphology deals with the critical analysis of the morphological processes and their products, the assemblage of landforms within a distinct time span. Environmental Geomorphology generally defines the scientific study of morphological process and landforms with respect to nature.

The nature of the discipline relates principally with physical features of earth history. The subject principles are the mirror of face-features exposed over earth surface. The processes and landforms and the so-called natural embalance are thought to be deeply influenced by human interactions. Thus, the eco-culture cropped over this planet is closely corresponded with physico-cultural phenomena. The growing human races have designed an unique glimpse of cultural landscape and their creativity have gifted him all necessities in possibilities. In the techno-scientific era, he is making race towards space. He has destructed the nature for some to the fact that 'Lord Rama' chose Chitrakut to pass his time in exile as a lap of grandeur, peace and abundant natural wealth. Many saints and sages also found the place praiseworthy and ideal for transcedental meditation etc. 'Goswami Tulsidas' also could get a 'darshan' of 'Lord Rama' at this very spot.

The evidences mirrored in 'Ramayana' also depict that the 'Chitrakut Dham' is not merely confined in plateau region but it has more centimental touch towards north fertile Yamuna plain also. From Rajapur (the birth place of Tulsidas situated along the bank of river Yamuna in north) to Gupt-Godavari (south-western part of scarp zones) the entire region has become scared area and dotted with various pilgrimage centres of the 'Chitrakut Dham'. Paisuni river which is locally termed as 'Mandakini' contains many religious spots along its course. Thus, it may be marked that the 'Tirthas' of 'Chitrakut Dham' are either located along the river courses or along the scarpzones where typical landscapes have developed. The typical geomorphological features as well as the rich cultural milieu are a source of great attraction for the tourists and the researchers in various disciplines.

History reveals that the region was virtually a cockpit of constant warfare among the regional power like the Bundela rulers from the beginning of the 13th to the end of the 18th century. Due to warfare, the agriculture was totally neglected and scarcities and famines of that time occasionally deepened the crisis. Various projects and land reform measures were launched by the Chandels and later by the Britishers to improve the region. After Independence, the extension of agricultural practices have almost removed the original forest cover from many parts of the region. The rich alluvial plain has degraded due to gullying and sheet erosion. The defective cropping pattern, the lack of irrigational facilities, transport and communications, the poorly developed 'Patha soil', accumulation of sand in agricultural fields of the Piedmont zone, stone quarrying etc. are the main causes for the underdeveloped economy of the region. Obviously, it has attracted the attention of the Government as well as individual researchers and attempts are being made to improve the economy of the region and to mountain the ecological balances.

The present study encompasses the above theme and the work is divided in ten chapters.

Chapter 1 deals with the environmental outlook and research methodology. A conceptual framework of eco-system, form and process, process form chronology and eco-management schemes are discussed in the first phase. The interpretation of physio-cultural eco-system is mirrored through the study of topographical sheets, survey techniques and illustrated by suitable cartographic methods. In addition the research design, study orientation and its importance, precedence and performance of eco-prospects and the proposed research plan have also been incorporated.

Background of major morphometric properties is discussed in Chapter 2. In this context, linear, areal and relief aspects of drainage network have been analysed at basinal (selecting 20 drainage basins of the region) and regional level. The appropriate mathematical techniques and principles proposed in this regards have also been tested to arrive at the final results. In the second phase of morphometric interpretation, the major components of drainage hierarchy like drainage density, drainage density ratio, drainage texture, stream frequency, average slope, clinographic analysis and drainage dissection have been taken into account which form a base for further discussion dealt with in the succeeding chapters.

Chapter 3 highlights the conceptual framework of process-response and process types along with the measurement techniques of significant processes. The intensity of various morphological processes has been determined on the basis of the Peltier's and Wilson's models.

The detailed study of weathering processes and mass movement (Chapter 4) is carried out through review work of previous literatures and with the help of extensive field survey of the region. Physical, chemical and biotic weathering are marked with several examples in different litho-units. Mass movement and its associated processes like landslides, rockfalls, slums and surface wash are also interpreted.

An introduction to erosional typology, techniques, chief determinants, system and mechanism of fluvial processes

(Chapter 5) are the major task of the study applied in fluvial processes. Infiltration, overlandflow through flow, saturated overland flow, pipe flow, surface and sub-surface flow, sheet wash, rill wash and gullying have been discussed and critically examined in the region. The chapters 6 and 7 are concerned with weathering, the associated landforms assemblage (Chapter 6) and the fluvial topography (Chapter 7), while in Chapter 8, the micro-analysis of some major landforms like caves and tors are included.

The evolutions of landforms due to weathering processes are analysed under the processes involved viz. constant volume weathering, expansion weathering, differential weathering, stripping and karsting. Landforms caused by unloading and sculping are marked while some peculiar examples of spheroidally weathered blocks, colour banding, unloading domes, sinkholes, caves and blind valleys are also noted.

Chapter 7 illustrates the nature and characteristics of fluvial topography in the region. General cyclic stages of presently topography is discussed in the first part of this chapter. The study of the development of slopes and scarps and some other erosional features of valley development, development of waterfalls, rapids, springs and seepages and depositional landforms have also taken into consideration in the later part of the discussion.

The morphological survey and laboratory analysis of Gupt-Godavari caves and the evolution of granitic tors have been critically analysed in Chapter 8. Chemical, geochemical, petrographic analysis and trace element content of dolomite rocks have been obtained on the basis of the rock samples and laboratory analysis. The morphological plan of the caves was prepared by prismatic compass survey. Tors were identified with the help of topographical sheets and studied extensively through field observations.

Environmental problems and eco-management planning have been discussed in the last two chapters (Chapter 9 and 10). Chapter 9 highlights the major environmental problems causing degradation and hazard of the region. The systematic critical review of major eco-problems like deforestation, hazard by stone quarries, water scarcity and soil moisture deficit, soil erosion, ravination and the associated natural hazard have been analysed.

The last chapter (10) presents the morphological plan and eco-management planning of the region.

The book includes 68 figures and 37 tables. The cartographic representation is based on topographical sheets interpretation and the mathematical analysis.

Govind Prasad

Contents

1

Environmental Outlook and Research Methodology

PROFILE OF ECO-CIRCUMFERENCE

Eco-circumference or the term environment has been defined as the sum total of all conditions and influences that affect the development and life or organisms. This is a comprehensive definition as it stresses its totality and every living organism, from the lowest to the highest including human being, has its own environment. In a wide term, it includes every action, reaction, interrelationship, response factors, orientations and reorientations of mass, energy modifications and transformations in a number of ways that set a balance among them at least for some time, the said balance being adjusted and readjusted by the forces of dynamism; natural or cultural. The earth surface presents an ever-changing 'geocomplex' which comprises the physio-system i.e. the cycle of abiotic components of the landscape with their dynamic relationship and also the biotic complexes (better known as biocenosis) working on the said landscape. The nature of the action under eco-circumference is termed as 'holocoenosis'. It pertains to those factors of the structure of function of natural systems which exist as a vast complex and therefore do not act separately and independently. This principle lies at the core of ecological thinking (Jayaswal, S.N.P. and Prasad, G. 1989). Environment is an inseparable whole and is constituted by the interacting system of physical, biological and cultural elements (Fig. 1.1) which are interlinked individually as well as collectively in myriad ways.

Physical elements (space, landforms, water bodies, climate, soils, rocks and minerals) determine the variable character of human habitat, its opportunities as well as limitations. Biological elements (plants, animals, micro-organisms and man) constitute the biosphere. Cultural elements (economic, social and political) are essentially man-made features which go into the making of cultural milieu (Singh, S. and Dubey, A. 1983). Thus, it may be expressed that all surroundings of a designated eco-system is environment.

The Webster's Dictionary has aptly defined ecology as "the totality or pattern of relations between organisms and their environment." The word ecology is derived from the Greek word 'Oikos' meaning 'home' or place to live. Literally ecology is the study of organisms 'at home' and is defined as the study of the relation of organisms or groups of organism to their environment.

Following to Odum, E.P. (1971) modern ecology can be considered in terms of the concepts of 'level of organisation' visualized as a sort of biological spectrum as depicted in Fig. 1.2. Community, population organism, organ, cell and gene are widely used terms for several major biotic level illustrated in the hierarchical arrangement from large to small in Fig. 1.2. Interaction with the physical environment (energy and matter) at each level produces characteristic functional systems. Thus the circumference of ecology and environment is most common.

Generally, ecology has been used to measure inter-relationship of various organisms with their changing environments. This term is first used by Emst Haechal in 1866. The scientific terms 'geoecology', 'synecology' and 'biocenosis' have taken birth in eco-circumsference. The term 'Geoecology' also known as 'Landscape Ecology' was coined by Carl Troll in 1939 to refer to the qualitative and quantitative interactions between different components of the geocomplex.

The application of ecology to the functional aspects of biotic communities is the field of 'Synecology'. The self regulating and self sustaining community of plants and animals has come sort of stable equilibrium with the environment which is known as 'Biocenosis'.

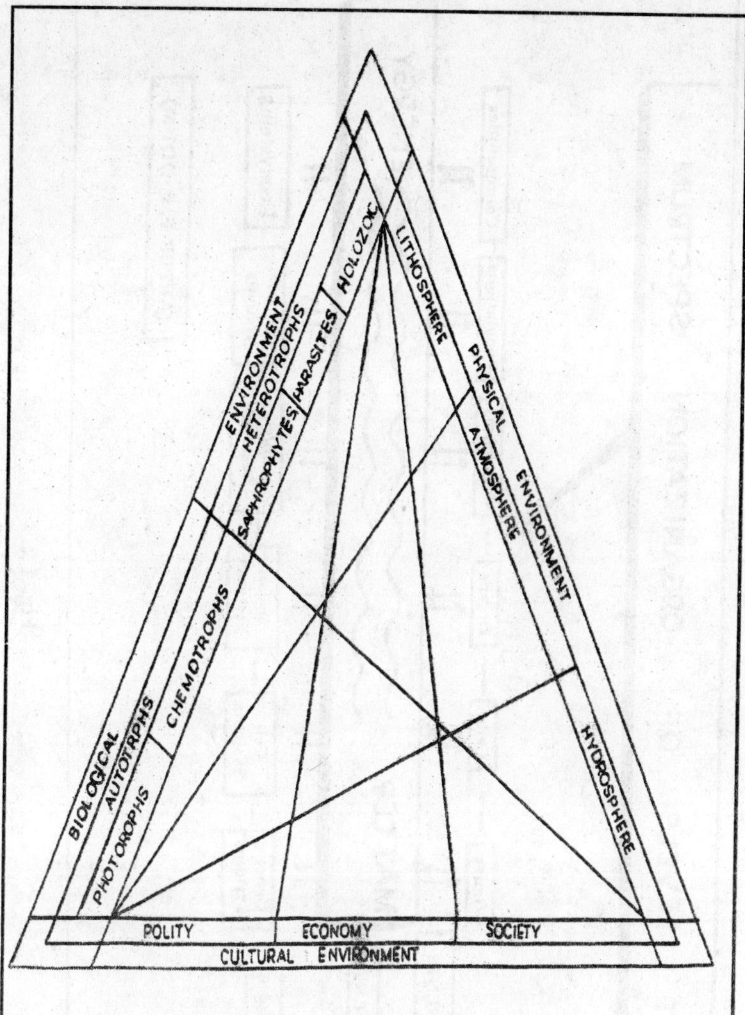

Fig. 1.1: Interaction of physical, biological and cultural elements of environments

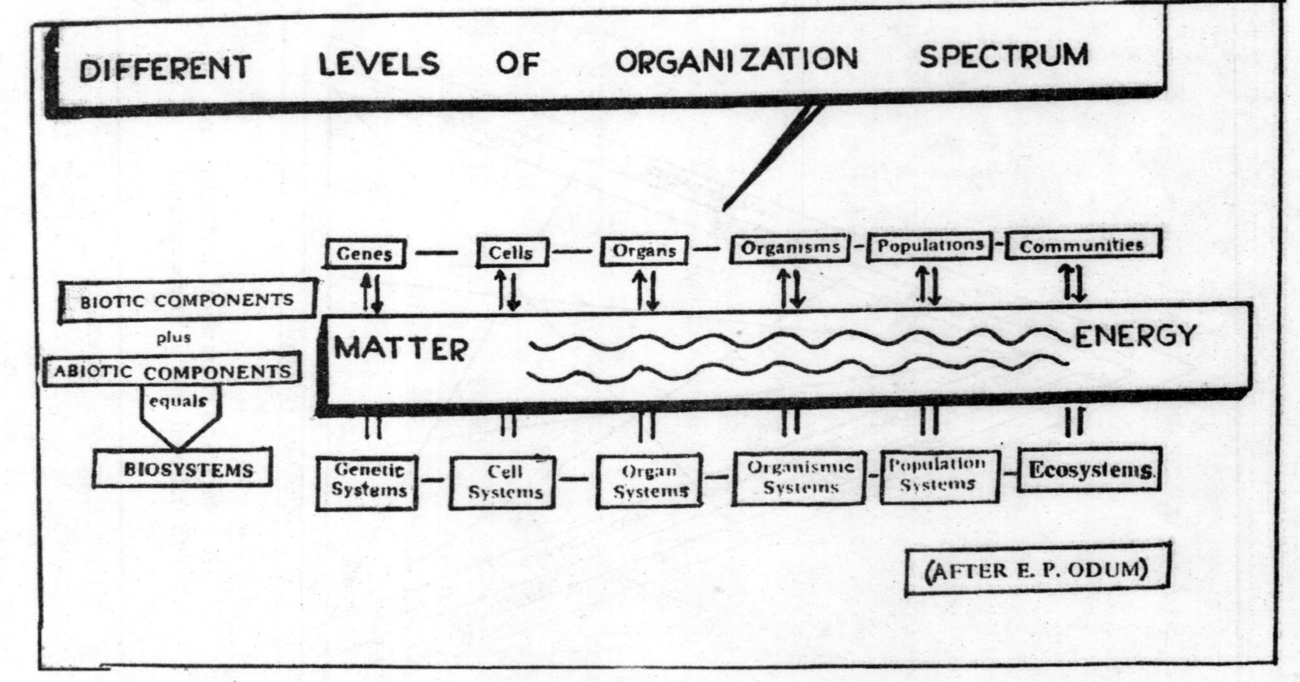
DIFFERENT LEVELS OF ORGANIZATION SPECTRUM
BIOTIC COMPONENTS
plus
ABIOTIC COMPONENTS
equals
BIOSYSTEMS
Genes
Cells
Organs
Organisms
Populations
Communities
MATTER
ENERGY
Genetic Systems
Cell Systems
Organ Systems
Organismic Systems
Population Systems
Ecosystems
(AFTER E. P. ODUM)

Fig. 1.2

The concept of eco-balance is directly related to man's transformatory action in the geosphere i.e. extension of noosphere (the reasoning zone) through man's activity. The other sub-zones in different perspectives may be called as sociosphere and technosphere which respectively promote humanized and technocratic (cultural) eco-society. In the scientific development a new branch called 'Habitat Ecology' has also been postulated with respect to comprehend 'Habitat Science.'

ECO-SYSTEM, FORM AND PROCESSES: A CONCEPTUAL OUTLOOK

Eco-system is an ecologic system composed of an organic community (Biotic complex) of plants and animals (Biome) viewed within its physical environment or habitat. It is essentially a somewhat more technical term for 'a segment of nature' and the result of interaction between biological, geochemical and geophysical systems. It is often used in 'Ecology' for the physical background. The study of an 'e' provides a methodological basis for complex synthesis between organisms and their environment.

'Eco-system' was the term enunciated by 'Tansley' to express 'interacting system' comprising living things and their non-living environment (Transley, A.G. 1935). Stoddart, D.L. (1965) has developed it as the fundamental organising concept of geography because "Firstly, it is monistic, it brings together environment, man and the plant and animal worlds within a single framework, within which the interaction between the components can be analysed...Secondly, eco-systems are structures in a more or less orderly, rational and comprehensible way. The essential fact here, for geography, is that once structures are recognised they may be investigated and studied. Thirdly, eco-systems function, they involve continuous through-put of matter and energy. Fourthly, the eco-system is a type of general system and possesses the attributes of general systems. In general, system terms the eco-system in an open system tending toward a steady state under the laws of open system thermodynamics. Eco-system is an integrated system in nature which may be studied as an independent entity, e.g. a rolling log in the forest, a coral atoll, a continent or the earth with all its biota. Eco-system ecology deals with plant and animal communities in terms of total eco-system. From a structural standpoint four constituents of eco-system can be recognised (Fig. 1.3). Fig. 1.3 denotes (i) abiotic substances, basic elements and compounds of the environment; (ii) procedures, the autotrophic organisms, largely the green plants; (iii) the large consumers or macroconsumers, heterotrophic organisms, chiefly animals that ingest other organisms or particulate organic matter; (iv) the decomposers or microconsumers (also called saprobes or saprophytes) heterotrophic organisms, chiefly the bacteria and fungi that breakdown the complex compounds of dead protoplasm, absorb some of the decomposition products and release simple substances usable by the producers.

The clearer perception of eco-system can be obtained by Fig. 1.4. Fig. 1.4A is a combination of energy and nutrients (part of exchangeable matter) flow in eco-system. Fig. is the expansion of simplified energy flow diagram showing three tropics levels (boxes numbered 1, 2, 3) in a linear food chain. Standard notations for successive energy flows are as follows: 1 = total energy input; L_A = light observed by plant cover; P_G = gross primary production; A = total assimilation; Pn = net primary production; P = secondary (consumer) production; Nu = energy not used; N_A = energy not assimilated by consumers; R = respiration (Odum, E.P. 1963). Fig. 1.4C is a nutrient cycle based on leaves which fall into shallow esturine water. Leaf fragments acted on by saprotrophs and colonised by algae are eaten and reeaten (coprophagy) by a key group of small detritus consumers which in turn provide the main food for game fish, herom, stork etc. (Odum, E.P. 1956). Park, C.C. (1980) advocated that the eco-system stresses the unity of all parts of the environment including both biotic and abiotic elements associated with morphological processes; it facilitates measurement and comparison of the components of different types of ecological communities and identification of equilibrium and non-equilibrium states; and it highlights basic ecological principles which should be carefully evaluated when guidelines of ecological management and exploitation are completed:

Law one : Everything is connected in everything else.

Law two : Everything must go somewhere.

Law three : Nature knows best.

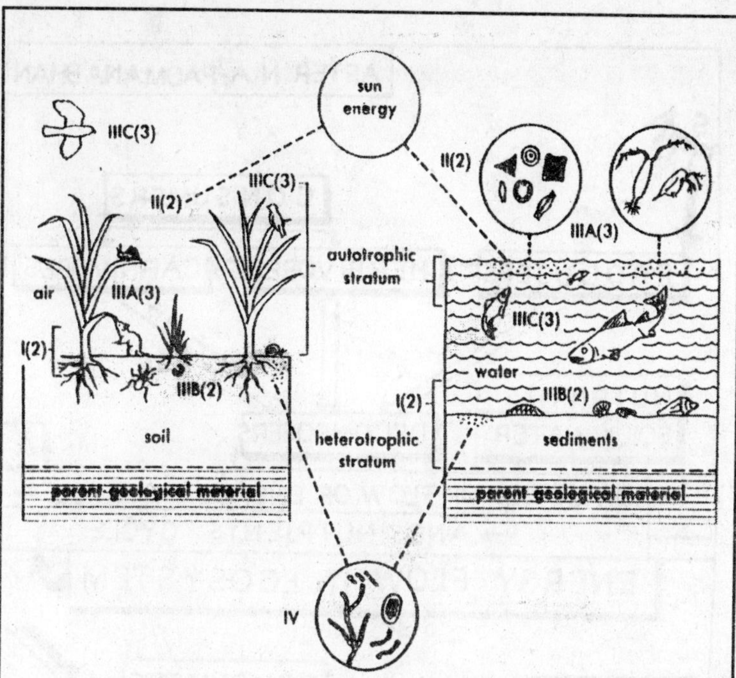

Fig. 1.3: Comparison of the gross structure of a terrestrial ecosystem (a grassland) and an open-water ecosystem (either fresh water or marine). Necessary units for function are: I: Abiotic substances (basic inorganic and organic compounds). II: Producers (vegetation on land, phytoplankton in water). III: Macroconsumers or animals: (A) direct or grazing herbivores (grasshoppers, meadowmice, etc. on land; zooplankton, etc. in water); (B) indirect or detritus-feeding consumers or saprovores (soil invertebrates on land; bottom invertebrates in water); (C) the "top" carnivores (hawks and large fish), IV: Decomposers, bacteria and fungi of decay.

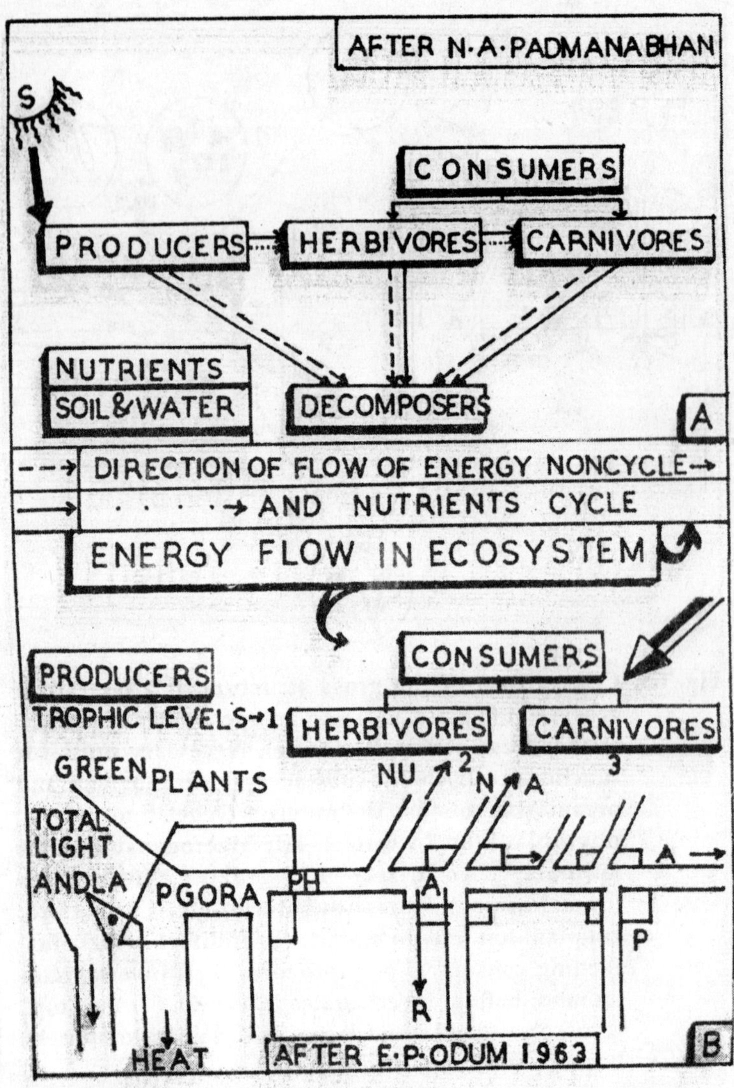
AFTER N·A·PADMANABHAN
S
CONSUMERS
PRODUCERS
HERBIVORES
CARNIVORES
NUTRIENTS
SOIL & WATER
DECOMPOSERS
A
DIRECTION OF FLOW OF ENERGY NONCYCLE
AND NUTRIENTS CYCLE
ENERGY FLOW IN ECOSYSTEM
CONSUMERS
PRODUCERS
TROPHIC LEVELS→1
HERBIVORES
CARNIVORES
2
3
NU
N
A
GREEN PLANTS
TOTAL LIGHT
ANDLA
PGORA
PH
A
A
P
R
HEAT
AFTER E·P·ODUM 1963
B

Fig. 1.4

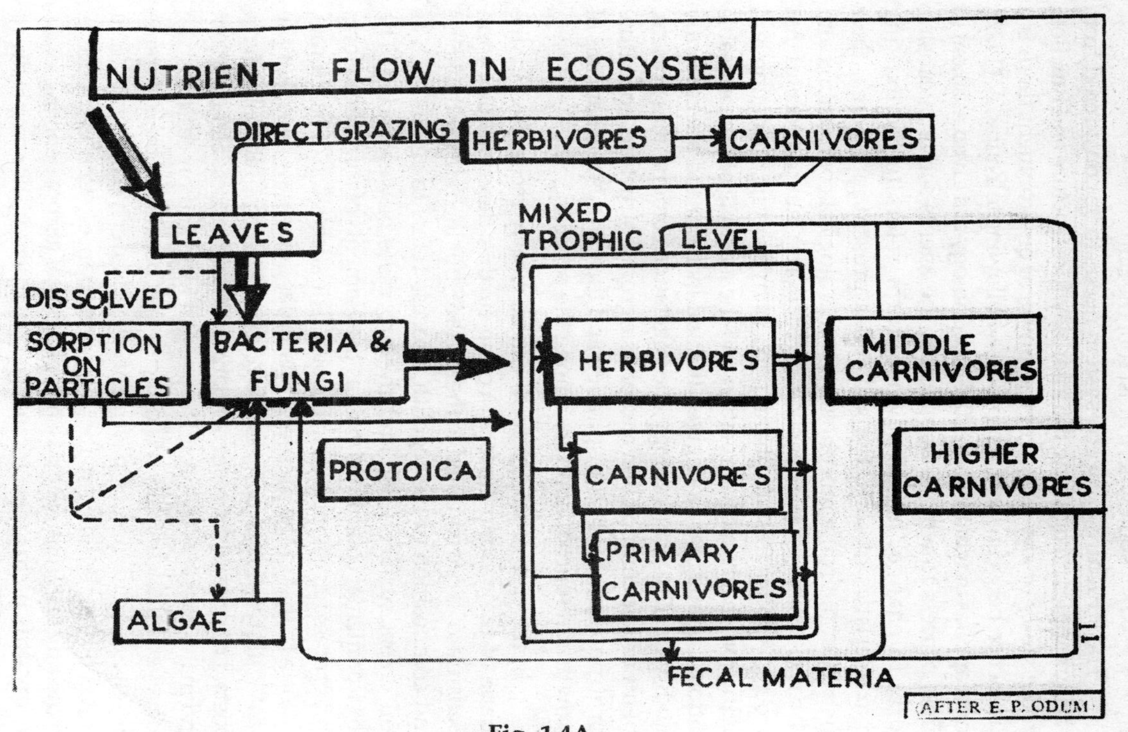
NUTRIENT FLOW IN ECOSYSTEM
DIRECT GRAZING
HERBIVORES
CARNIVORES
LEAVES
MIXED TROPHIC LEVEL
DISSOLVED
SORPTION ON PARTICLES
BACTERIA & FUNGI
HERBIVORES
MIDDLE CARNIVORES
PROTOICA
CARNIVORES
HIGHER CARNIVORES
PRIMARY CARNIVORES
ALGAE
FECAL MATERIA
(AFTER E. P. ODUM

Fig. 1.4A

Law four : There's no such thing as a free lunch (somebody somewhere must foot the bill).

The structure of an eco-system needs to be considered from various angles recognising the interplay for structure and function. In the study of any aspect of the physical environment there is a distinction between form and process. Following to Young, A. (1972), form applies to what is there, the morphology at a given moment in time; process to what is happening, the agents active in causing form to change. In soil science, form refers to the existing characteristics of soils, the soil types present in an area, their distribution, profile morphology, mineralogical composition and also non-visible properties such as reaction. Young advocates that process applies to such phenomena as hydrolysis, oxidation and reduction, mechanical eluviation and the leaching of exchangeable cations, agents that have given the soil its form and are currently modifying it. In plant ecology, the vegetation communities present, their floristic composition and physiogromic structure, constitute form, photosynthesis transpiration and the various mechanisms of growth and reproduction are among the process.

In the case of slopes, form can be clearly recognised by the shape of the ground surface. Young further describes that form 'Sensu lato' is not confined to the surface, but include the thickness and composition of the regolith. Process refers to agents such as soil creep, surface wash and the processes of weathering. The environmental conditions exist as the elements of form. Both form and process have appeared in the past. The succession of past forms, leading to that of the present, constitutes some new ones. For the past with respect to process, no recognised comprehensive term exists. The general problem of relating form to process at the present and in the past is as follows. The present form and processes can be surveyed but technically it is difficult task because of their extreme slowness. Evidence of past processes is derived mainly from deposited materials but the interpretation of such evidence is frequently uncertain. The past morphological history denoting forms is very difficult to judge. It can be reconstructed. The complexity of processes, the slowness of form change, the spatial diversity of the environmental conditions and their variation in time are among the difficulties in the way of achieving an understanding of the relations between form and their evolution.

PROCESS-FORM CHRONOLOGY AND ECO-MANAGEMENT SCHEMES

The study of form and process is between the descriptive and genetic approaches to the environment. The geomorphic laws and the studies of processes were first enunciated by Hutton in 1785. It was beautifully restated by Playfair in 1802 and popularized by Lyell in the numerous editions of his '*Principles of Geology*' explaining the concept that "the same physical processes and laws that operate today operated throughout geologic time, although not necessarily always with the same intensity as now." Hutton advocated that 'the present is the key to the past.' He applied this principle very rigidly and postulated that geologic processes operated throughout geologic time with the same intensity as now. But the concept of Hutton is not possible in nature because numerous examples could be cited to show that the intensity of various geologic processes has varied through geologic time, but there is no reason to believe that streams did not cut valleys in the past as they do now.

The importance of form and processes can be well recognised by Davis 'Trio' structure, process and stage. In his 'Geographical cycle' he stated that "Landform is a function of structure, process and stage. The concept of Davis is validated till now. It can be examined that the geologic structure plays dominant role in the evolution of landforms as a chief controlling factor.

One can examined a different types of relief features and a multidimensional landform assemblage over the surface of the earth because the geomorphic processes operate at differential rates.

In the study of fundamental concepts of geomorphology Thornbury, W.D. (1954) postulated that geomorphic processes leave their distinctive imprint upon landforms and each geomorphic process develops its own characteristic assemblage of landforms. The observation of Thornbury is found correct in the field of analysis. Landforms have their individual distinguishing features depending upon the geomorphic process responsible for their development. Landscapes are the products of a group of processes. It is observed that the different erosional agents acting upon the earth's surface produced an orderly sequence of landforms. The present day multicyclic landforms are caused by the actions of a complex set of geomorphic processes.

Eco-degradation is a major problem arising before the existing civilization of the earth surface. It is a matter of great satisfaction that India took note of all these integrated environmental problems and this concern was the first time articulated in the Fourth Five-Year Plan (1969-74). The plan drew our attention to the environmental issues in these words. It is an obligation of each generation to maintain the productive capacity of land, water, air and wildlife in a manner which leaves its successors some choice in the creation of a healthy environment...Planning for harmonious development recognizes this unity of man and nature. Such planning is possible only on the basis of comprehensive appraisal of environmental issues particularly economic and ecological. There are instances in which timely specialized advice on environmental aspects could have helped in project design and in averting subsequent adverse effects on environment, leading to loss of invested resources. It is necessary, therefore, to introduce the environment aspect into our planning and development. The Sixth Five-Year Plan (1980-85) attached great importance to the protection of environment initiating the integrity of our natural resources and soil water forests, wildlife etc. The Seventh Five-Year Plan was operated in the right direction of all development of environment. The Ganga Authority for cleaning the water of the Ganges and the surrounding slums has been established.

The fundamental issues of eco-development has been meritoriously getting reflections in various national and international conferences such as those of Finland (1971), Stockholm (1972) and Vancouver (1976) convened under the auspices of the United Nations. The Stockholm conference (June 1972) on Human Environment was held realising the importance of environmental pollution problems by a scientific advisory committee of United Nations. The National Committee on Environmental Planning (NCEP) reconstituted in April 1981 has done valuable work in a number of areas related to eco-development planning. The Government of India constituted a High Power Committee under the Chairmanship of the Deputy Chairman of the Planning Commission, Shri N.D. Tiwari. It was submitted to the Prime Minister in September 1980 and had identified the following major areas of environmental concern:

1. Environmental pollution;
2. Mismanagement of land and water resources;
3. Depletion of natural resources consciously and in ignorance;
4. Poor condition of human settlements; and
5. Need for environmental awareness and education.

The committee suggested some of the important legislative measures concerning with biosphere reserves, protection of grazing lands, protection of endangered species, toxic substances control act, scientific landuse, prevention of noise, pollution and the prevention of denudation of forests. The National Environmental Advisory Committee was also constituted in 1983 for highlighting environment issues and giving advice on remedial action. A new integrated department called 'Department of Environment, Forests and Wildlife' in the Ministry of Environment and Forests came into being in 1985. The allocation of work to the new department includes National Landuse and Wasteland Development Council, National Wasteland Development Board, Central Ganga Authority (CGA) set up in February 1985 in the Department of Environment. The CGA overseas the implementation of the Ganga Action Plan. Several actions like pollution control, land resource management, conservation of natural living resources, environmental awareness programmes, environmental impact assessment and monitoring and environmental information are set in motion to preserve the environment.

There is a specific reference in our constitution about environment in the Directive Principles of the State policy. The following duties have been held been laid down for the state and the citizen.

– **Article 48A** (42nd Amendment 1976)

"The state shall endeavour to protect and improve the environment and to safeguard the forests and wildlife of the country."

– **Article 51A** among other things states

"It shall be the duty of every citizen of India— (I) to protect and improve the natural environment including forests, lakes, rivers and wildlife and to have compassion for living creatures."

There are over 400 central and state legislations which may be relevant to varying extent to environmental protection. Our laws relating to environmental problems are scattered over different statute books. In recent years our Parliament has enacted the following legislation to save the environment from being polluted:

1. The Water (Prevention and Control of Pollution) Act, 1974;
2. The Water (Prevention and Control of Pollution) Cess Act, 1977;
3. The Air (Prevention and Control of Pollution) Act, 1981;
4. The Wildlife (Protection) Act, 1972;
5. The Forest (Conservation) Act 1980; and
6. The Environmental (Protection) Act 1986.

Some of the urgent environmental aspects requiring proper legal backup and/or updating of existing legislations are:

1. Landuse Planning (including cropland, grassland, woodland);
2. Revision of Forest Policy and corresponding updating to the Forest Act;
3. Environmental Impact Assessment (EIA);
4. Catchment Areas Protection;
5. Crazing and Grasslands;
6. Revision of the Water Act; 1974;
7. Biosphere Reserves;
8. Revision of Mines and Minerals (Regulation and Development) Act; 1957;
9. Hazardous Substances Management;
10. Resource Recovery, Recycling and Reuse;
11. Indian legislation for effective implementation of Convention on International Trade in Endangered Species and Wild Fauna and Flora (CITES); and
12. Marine Resources Conservation (including over-fishing, estuaries, mangroves, control of pollution in coastal waters and EEZ etc.)

LOCATION AND DELIMITATION OF THE STUDY AREA

The region, popularity known as 'Chitrakut Dham' lies between 24° 56′ 15″ N latitude and 25° 27′ 52″ N latitude 80° 40′ E longitude and 80° 14′ 38″ E longitude. It occupies a fairly large area (1851.402 sq. km.) in the south-eastern part of Banda district in Uttar Pradesh and north-eastern part of Satna district in Madhya Pradesh (Fig. 1.5). The region comprises the greater part of six development blocks viz. Pahari, Ram Nagar, Naraini, Shivrampur, and Manikpur of Banda district (U.P.) and Majhgawan of Stana district (M.P.). It is a holy centre of 'Lord Rama' according to Hindu mythology.

The evidences mirrored in 'Ramayana' depict that the 'Chitrakut Dham' is not merely confined in plateau region but it has its more centimental tough towards north fertile Yamuna plain. From Rajapur (as located along the bank of river Yamuna in the north and the birth place of Goswami Tulsidas) to Gupt-Godavari (near Chaubepur village situated in the south-west along the scarp zone) the sacred passage is marked with various holy centres of 'Chitrakut Dham'. Paisuni which is termed here as 'Mandakini' river also determines several holy centres like Sitapur, Purani Lanka, Kotra Gudar, Akshyawat, Pramod ban, Janki kund, Sirsa ban, Phatakshila, Sati Anisuiya Ashram, etc. from north to south respectively. In north-east and south-west direction, Bankesidh (denoting cave and spring conditions), Kottirath, Pampeshwar, Devangana, Narsingh Dhara, Sita Rasoi, Hanuman Dhara, Sidh Baba, Modhwa (cave and spring) and Gupt-Godavari are the most important 'tirth' of the 'Dham' which cover the total length of scarp faces. Thus the region is delimited between Banganga, Paisuni, Yamuna, Ganta interfluvial zones. It is true in religious content that Paisuni possesses its more religious importance as it drains slowly towards Yamuna river in the north. It is also remarkable that the 'tirthas' of 'Chitrakut Dham' are either located along the river courses or along the scarpzones where typical landscapes are noticed. Thus, they attract both ways e.g. one by their religious importance and other by their unique morphological features. The extension of the region is also governed by these two factors.

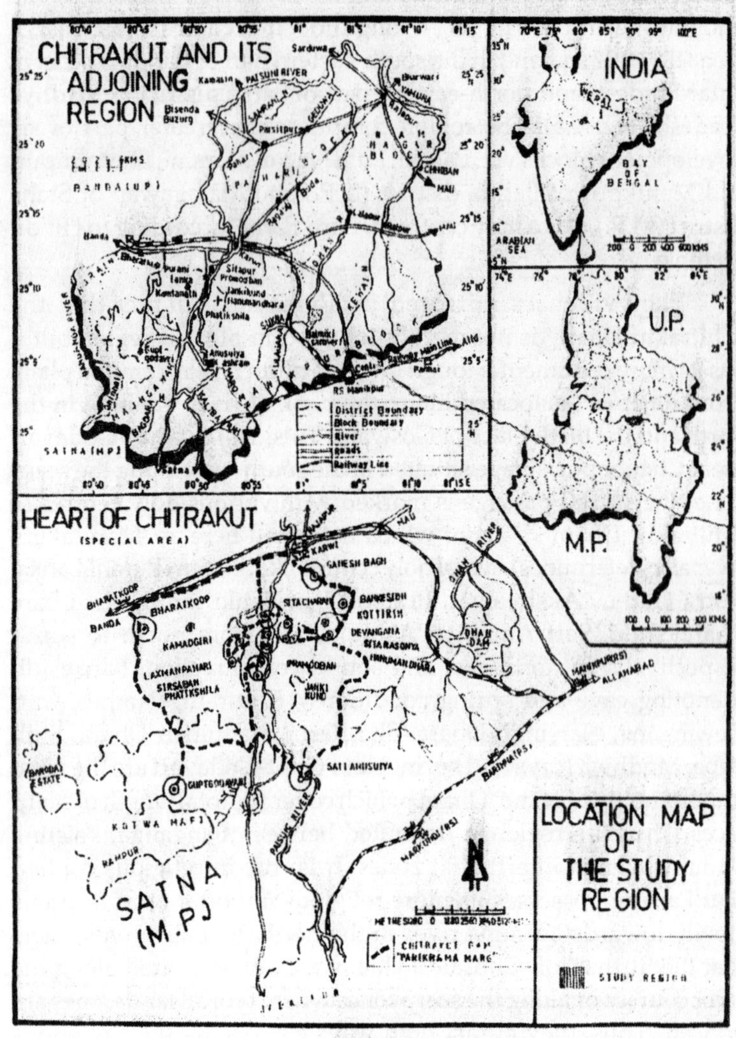
CHITRAKUT AND ITS
ADJOINING
REGION
INDIA
BAY OF BENGAL
ARABIAN SEA
U.P.
M.P.
HEART OF CHITRAKUT
(SPECIAL AREA)
District Boundary
Block Boundary
River
Roads
Railway Line
SATNA (M.P)
LOCATION MAP
OF
THE STUDY
REGION
STUDY REGION

Fig. 1.5

The Yamuna and its major tributary Paisuni (the mainstream of the region) forms its northern boundary whereas the southern boundary is demarcated by Paisuni river and Sarbhanga nala. The eastern limits are defined by Ganta and Jaiwanti rivers and the Banganga, a tributary of Baghain river makes its western boundary. The north-west corner is designed by meandering course of Paisuni river while the south-eastern part is characterized by almost a regular line of Vindhyan escarpment which runs parallel to Allahabad-Satna central railway line trending roughly in ENE-WSW direction.

The region corresponds with the rolling surface of Mau development block (Banda district) in the north-east, the stony bed and sloping ground of Manikpur (Banda district) in the south-east, Satna hills in the south Ajaigarh upland in the south-west, Banda plain (Alluvial plain sedimented by Baghain and Ken rivers) in the west and gently sloping alluvial plain (caused by the Yamuna river) of Manjhanpur (Allahabad district) in the north respectively.

PHYSIO-CULTURAL ECO-SYSTEM

Eco-system can be recognised as physical and cultural intercorrelated to each other. These two depict the changing pattern of multidimensional structure, process and stage. The forces of the system develop some new kind of elements through the processes. The eco-system of 'Chitrakut region' can be explained in the following manner.

Physical Eco-system

Physical eco-system means the sum of the total of natural things which exist through topo, litho, climo, bio and chrono functions. Thus, the study of relief features, geological formations, drainage, climatic conditions of soil, vegetation and ground water balance are included in this eco-system.

Topographical Characteristics

The topographic variations (Prasad, G. 1987[9]) are marked between the altitude of 80 m (along the Yamuna river) and 446 m as observed at Gerua Pahar in the south (Fig. 1.6). The highest elevation can be measured in the extreme south whereas the lowest elevation may be traced in northern lowland area along the river valleys. The northern lowland country is shown by sparse contour lines which indicate low altitudes of below 180 m. The southern and south-western parts register closely spaced contours denoting the ridge and ravine topography. It is also shaded by dissected hills and several granitic hillocks and boulders of various size and shape. In general, the surface irregularity decreases from south-west to north-east direction according to general slope followed by the major streams of the region. The gradient also decreases in the same direction.

Following the numerical analysis of the absolute altitude of different height groups and its percentage of the total surface area the general outlook of topographical characteristics can be visualised (Fig. 1.7 and Table 1.1). Fig. 1.7 and Table 1.1 remark the following zones of generalized reliefs in the region:

1. Low lying areas with ravinous river valley sides (upto the height of 140 m);
2. Hillock country with marked gullies (between 140 m - 220 m);
3. Hilly country showing scarp faces and steep valley sides (between 220 m - 380 m);
4. High mountainous country (above 380 m).

Table 1.1: Topographical characteristics

S. No.	*Height groups (in m)*	*Percentage of the total surface area*	*General topographical characteristics*	*Zonal characteristics*
1.	Below 100	3.61	Ravinous land	
2.	100-140	44.98	Lowland with ravines along the river courses	I (48.59%)
3.	140-180	19.19	Residual hills with numerous gullies	
4.	180-220	10.60	Residual hills with cliffing marked hills top	II (2.79%)
5.	220-260	7.50	Hills top and scarp faces	
6.	260-300	8.37	Flat top of hilly country	III (2.41%)
7.	300-340	4.38	Hilly country	
8.	340-380	1.16	Hilly country	
9.	380-420	0.11	High mountainous country	IV (0.2%)
10.	Above 420	0.10	High mountainous country	

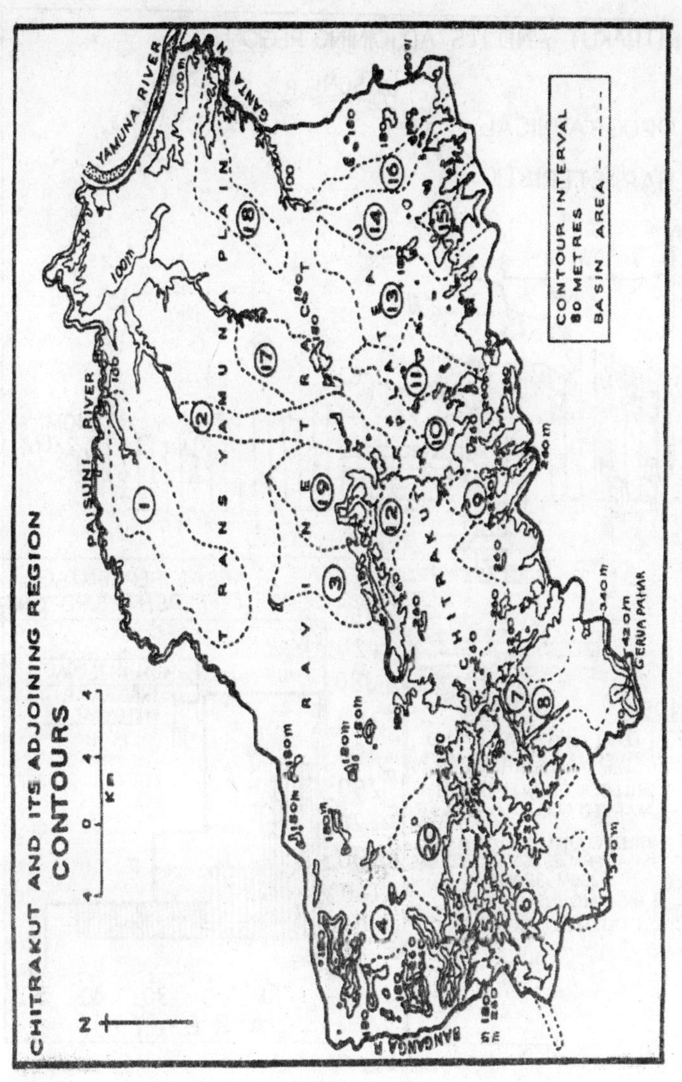
CHITRAKUT AND ITS ADJOINING REGION
CONTOURS
Km
YAMUNA RIVER
PAISUNI RIVER
GANTA
TRANS YAMUNA PLAIN
BANGANGA R
GERVA PAHAR
CONTOUR INTERVAL
80 METRES
BASIN AREA - - - - -

Fig. 1.6

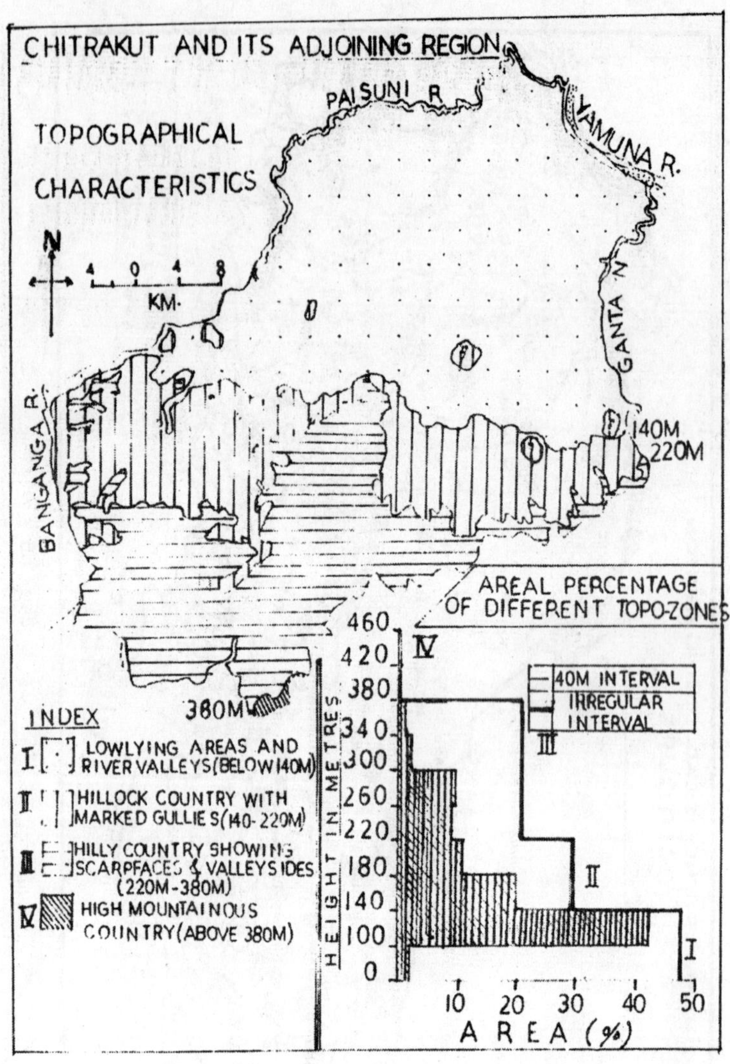

CHITRAKUT AND ITS ADJOINING REGION
TOPOGRAPHICAL
CHARACTERISTICS
PAISUNI R
YAMUNA R.
GANTA N
BANGANGA R.
N
4 0 4 8
KM.
140M
220M
380M
INDEX
I LOWLYING AREAS AND RIVER VALLEYS(BELOW 140M)
II HILLOCK COUNTRY WITH MARKED GULLIES(140-220M)
III HILLY COUNTRY SHOWING SCARPFACES & VALLEYSIDES (220M-380M)
IV HIGH MOUNTAINOUS COUNTRY(ABOVE 380M)
AREAL PERCENTAGE OF DIFFERENT TOPO-ZONES
HEIGHT IN METRES
460
420
380
340
300
260
220
180
140
100
0
IV
III
II
I
40M INTERVAL
IRREGULAR INTERVAL
10 20 30 40 50
AREA (%)

Fig. 1.7

Narrow cut deep valleys having steep valley-side slopes, deep, long and narrow gorges, parallel ridges with shoulders small and deep gullies, springs and waterfalls are some other topographic features of the region. A panoramic view of landscape characteristics relief dynamics and physiographic regions are also given in Figs. 1.8, 1.9, 1.10 and 1.11 respectively.

Yamuna plain, Ravine tract and the upland country cover 27.35 per cent (509.40 sq. km.), 41.52 per cent (768.20 sq. km.) and 30.97 per cent (573.20 sq. km.) of the total surface area of the region respectively.

Geological Formations

'Chitrakut Upland' is geologically prosperous because it contains the formations existed from the oldest Archeans to the recent alluvia. The country is mostly occupied by alluvium in the north and Vindhyan sandstone in the south (Fig. 1.12). The alluvial plain is thought to be underlain by medium to very coarse sand and gravel with clay, kanker and silt while 'Vindhyan sandstones' are presumed to be overlying by 'Tirohan limestone and Archean granites.' The Tirohan limestone are conspicuously absent over the scarpface to the east of Ohan river. It is interpreted to be due either to down faulting of the Kaimur plateau or to a northward recession of the original Vindhyan shoreline on the granite basement and later faulting thereby causing submergence and subsequent burial under the Pleistocene alluvial and lagoonal sediments of the areas. The Vindhyan forms the hilly tracts trending roughly in a ENE-WSW direction all along the country. The Bundelkhand granite occurs extensively south-western part of the country and around Naraini. The hillocks of the basement granite are occasionally capped by the Vindhyan rocks are surrounded by sub-recent to recent alluvial sediments which also occur to the west of Karwi, a few kilometres north of the scarpline.

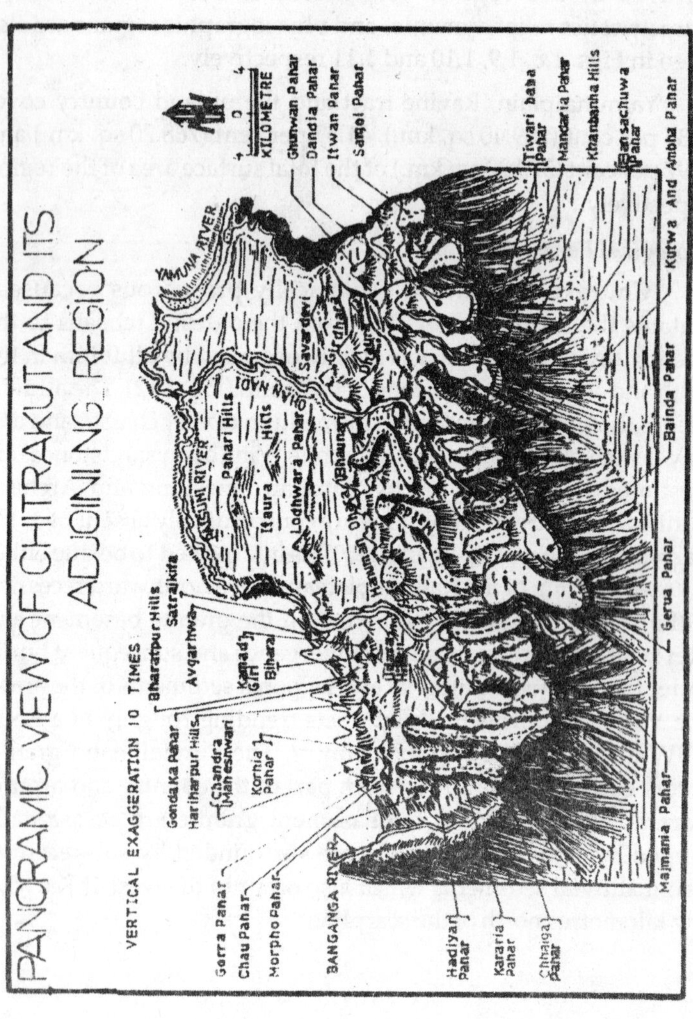
PANORAMIC VIEW OF CHITRAKUT AND ITS ADJOINING REGION
VERTICAL EXAGGERATION 10 TIMES
KILOMETRE
0
4
YAMUNA RIVER
PAISUNI RIVER
OHAN NADI
BANGANGA RIVER
Ghurwan Pahar
Dandia Pahar
Itwan Pahar
Samgel Pahar
Tiwari baba Pahar
Mandaria Pahar
Khanbanha Hills
Bansachuwa Pahar
Kutwa And Ambha Pahar
Bainda Pahar
Gerua Pahar
Majmania Pahar
Bharatpur Hills
Satrajkifa
Avgarhwa
Gondaka Pahar
Hariharpur Hills
Chandra Maheshwari
Kamad Giri
Bihara
Kornia Pahar
Pahari Hills
Itaura Hills
Lodhwara Pahar
Savardevi
Gorra Pahar
Chau Pahar
Morpho Pahar
Hadiyan Pahar
Kararia Pahar
Chhaiga Pahar

Fig. 1.8

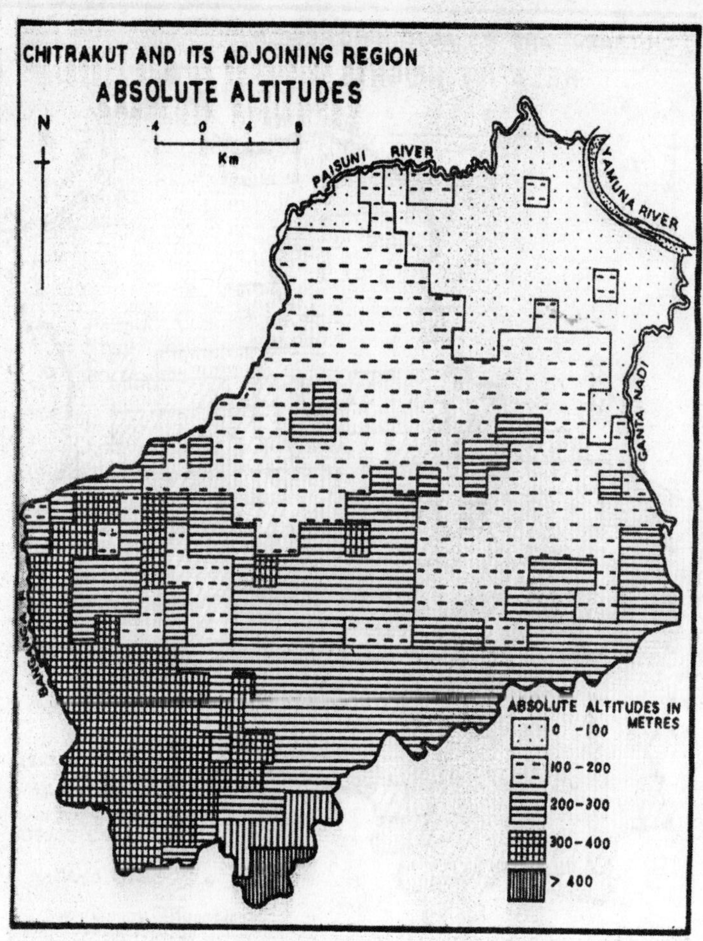
CHITRAKUT AND ITS ADJOINING REGION
ABSOLUTE ALTITUDES
N
4 0 4 8
Km
PAISUNI RIVER
YAMUNA RIVER
GANTA NADI
ABSOLUTE ALTITUDES IN METRES
0 -100
100 - 200
200-300
300-400
> 400

Fig. 1.9

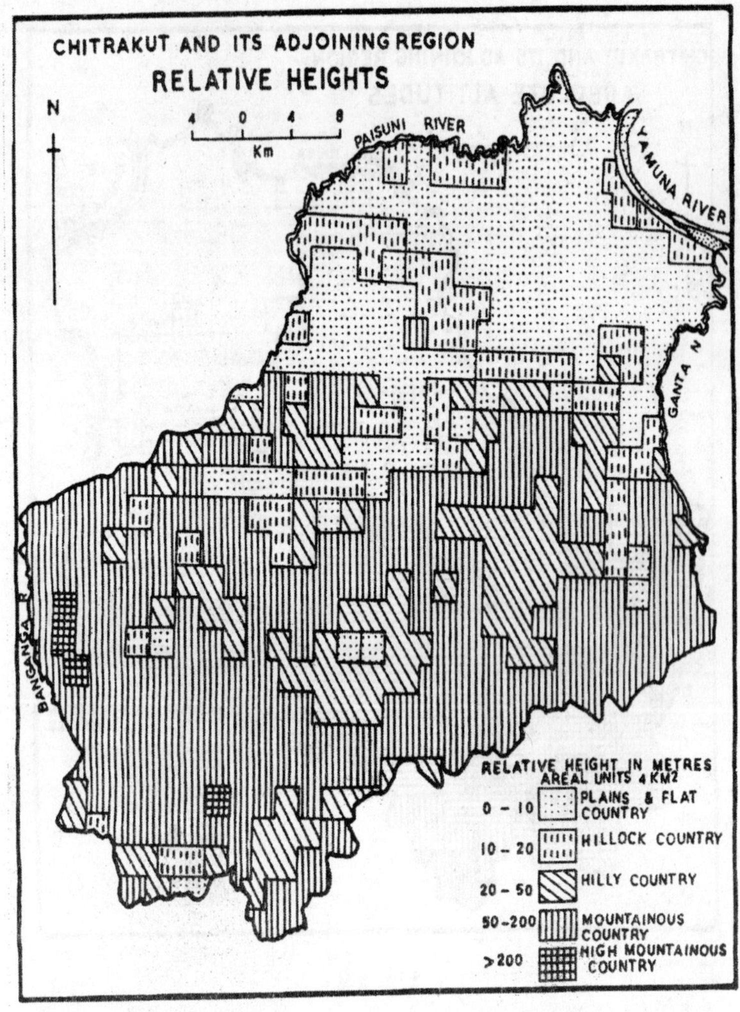
CHITRAKUT AND ITS ADJOINING REGION
RELATIVE HEIGHTS
N
4 0 4 8
Km
PAISUNI RIVER
YAMUNA RIVER
GANTA N
BANGANGA R
RELATIVE HEIGHT IN METRES
AREAL UNITS 4 KM2
0 - 10 PLAINS & FLAT COUNTRY
10 - 20 HILLOCK COUNTRY
20 - 50 HILLY COUNTRY
50-200 MOUNTAINOUS COUNTRY
>200 HIGH MOUNTAINOUS COUNTRY

Fig. 1.10

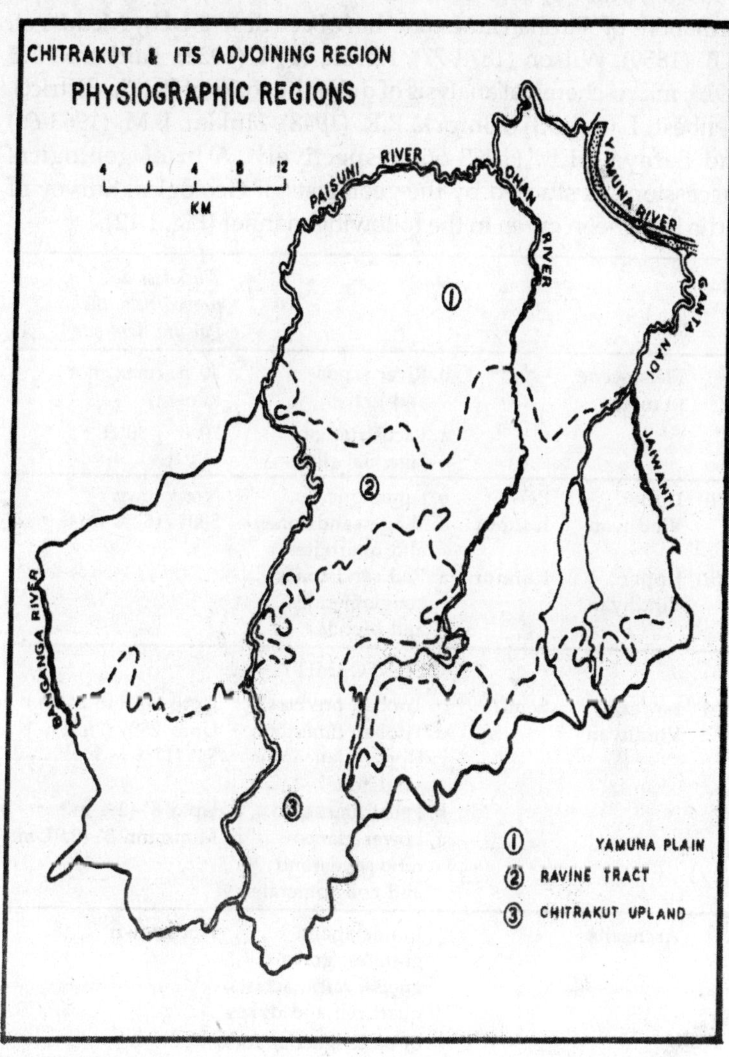
CHITRAKUT & ITS ADJOINING REGION
PHYSIOGRAPHIC REGIONS
4 0 4 8 12
KM
PAISUNI RIVER
OHAN RIVER
YAMUNA RIVER
GANTA NADI
JAIWANTI
BANGANGA RIVER
1 YAMUNA PLAIN
2 RAVINE TRACT
3 CHITRAKUT UPLAND

Fig. 1.11

The Vindhyan system has been studied mainly in the Son valley by some of the geologists including Meddlicot, H.B. (1860), Wilson (1889), Auden, J.B. (1933, Karwi area), Kedar Narain (1959-60, Karwi) and Hukku, B.M (1963-64) etc. The major formations of 'Tirohan limestone' have been described by Meddlicot, H.B. (1859), Wilson (1873-77), Heron, A.M. (1922), Fairbank, E.E (1925, micro-chemical analysis of dolomite rocks of Banda district), Henbest, L.G. (1931), Shrock, R.R. (1948), Hukku, B.M. (1963-66) and Safaya, H.L. (1963-66) respectively. A brief geological successions as studied by the geologists of Geological Survey of India have been given in the following manner (Fig. 1.12).

Age		*Series*	*Units*	*Thickness as proved near the Paisuni dam site*
C.	Pleistocene to recent	–	b. River sands and gravel	40 ft. (max. not known)
			a. Lacustrine or lagoonal silts	70 ft. - 80 ft.
B (ii)	Upper Vindhyan	Rewa	b. Sandstone	Not known
		Kaimur	a. White sandstones and quartzites	500′ (152.4 m)
B (ii)	Upper Vindhyan	Kaimur	a. Red sandstones conglomerates and breccia	–
			DISCONFORMITY	
B (i)	Lower Vindhyan	Semri	e. Tirohan breccia	Upto 50′ (15.24 m.)
			d. Tirohan limestone	Upto 250′ (76.2 m.)
			c. Upper glauconitic sandstone	58′ (17.6 m.)
			b. Pellet limestone	Upto 8′ (2.4 m.)
			a. Lower glauconitic sandstone and conglomerate	Minimum 3′ (0.91 m.)
A.	Archeans	–	Bundelkhand granites, gravels gneiss, mica, schists, quart relf and dykes of basic igneous rocks	Not known

Note: Formations are of very low dipping and almost horizontal.

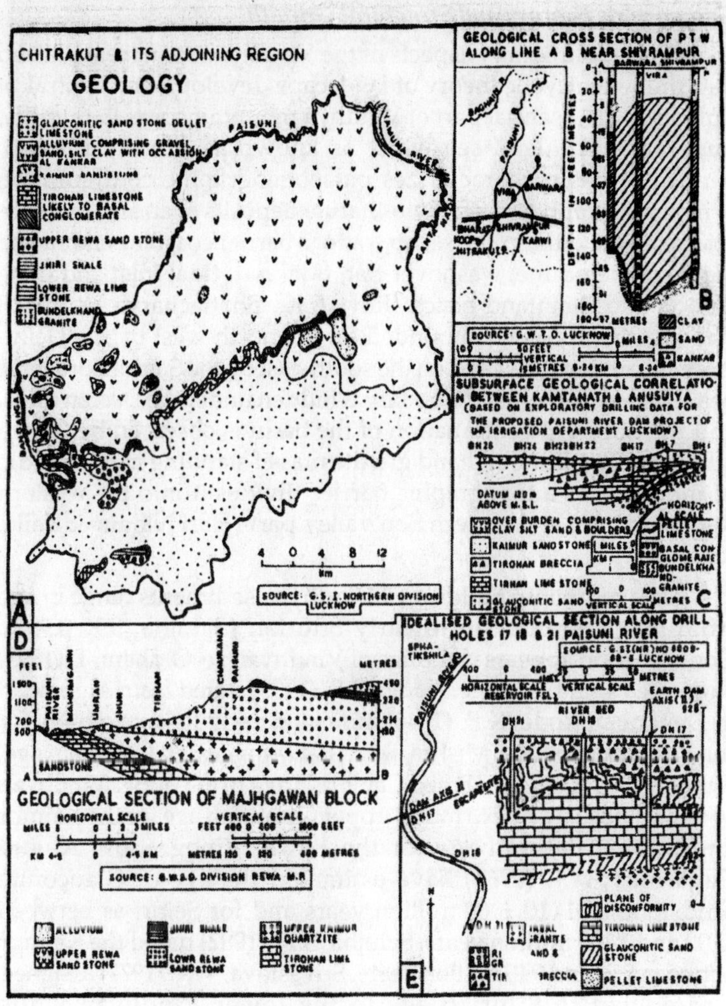
CHITRAKUT & ITS ADJOINING REGION
GEOLOGY
GLAUCONITIC SAND STONE PELLET LIMESTONE
ALLUVIUM COMPRISING GRAVEL SAND, SILT CLAY WITH OCCASSIONAL KANKAR
KAIMUR SANDSTONE
TIROHAN LIMESTONE LIKELY TO BASAL CONGLOMERATE
UPPER REWA SAND STONE
JHIRI SHALE
LOWER REWA LIME STONE
BUNDELKHAND GRANITE
PAISUNI R
YAMUNA RIVER
SOURCE: G.S.I. NORTHERN DIVISION LUCKNOW
A
GEOLOGICAL CROSS SECTION OF P T W ALONG LINE A B NEAR SHIVRAMPUR
SOURCE: G.W.T.O. LUCKNOW
CLAY
SAND
KANKAR
B
SUBSURFACE GEOLOGICAL CORRELATION BETWEEN KAMTANATH & ANUSUIYA
(BASED ON EXPLORATORY DRILLING DATA FOR THE PROPOSED PAISUNI RIVER DAM PROJECT OF U.P. IRRIGATION DEPARTMENT LUCKNOW)
DATUM 120 m ABOVE M.S.L.
OVER BURDEN COMPRISING CLAY SILT SAND & BOULDERS
KAIMUR SANDSTONE
TIROHAN BRECCIA
TIRHAN LIME STONE
GLAUCONITIC SAND STONE
PELLET LIMESTONE
BASAL CONGLOMERATE
BUNDELKHAND GRANITE
C
D
GEOLOGICAL SECTION OF MAJHGAWAN BLOCK
HORIZONTAL SCALE
VERTICAL SCALE
SOURCE: G.W.S.D. DIVISION REWA M.P.
ALLUVIUM
JHIRI SHALE
UPPER KAIMUR QUARTZITE
UPPER REWA SAND STONE
LOWR REWA SANDSTONE
TIROHAN LIME STONE
IDEALISED GEOLOGICAL SECTION ALONG DRILL HOLES 17 18 & 21 PAISUNI RIVER
PLANE OF DISCONFORMITY
TIROHAN LIME STONE
GLAUCONITIC SAND STONE
PELLET LIMESTONE
E

Fig. 1.12

Few details of surface and sub-surface geological formations including chemical analysis are also given in Fig. 1.13 and 1.14.

Geomorphic History and Structural Growth

The evolutionary aspects of the Vindhyan structural plateau determine the cyclic theory of landscape development. Chitrakut upland as an essential part of Vindhya mountains was the site of a huge ancient inland sea which is known as Vindhyan basin. Vindhyan basin characterizes palaeogeographic configuration principally composed of fluvio-marine deposits of an arid or a sub-arid region and the entire set up was of a barrier coastline embracing barrier beach dune, washover flat, tidal flat, tidal inlet and delta lagoon and mainland beach (Roy, A.K., Bhattacharya, A. 1982). The Vindhyan basin is bounded on the north-west by the Delhi-Aravalli orogenic belt and on the south-east by the Satpura orogenic belt. The south-western boundary is hidden under the Deccan traps, and the northern continuation of the basin is obscured by Ganga alluvium. The Bundelkhand granite massif standing in the middle of the basin as a topographic barrier divides it into the western Rajasthan part and eastern Son valley part (Hari Narain & Kaila, K.L., 1982).

The mountain building of Vindhyan sediments range in age from late Precambrian to early Silurian (Salujha, S.K. 1982). Contradiction appears in the age of Vindhyan i.e. Oldham, T. (1893) suggested Cambrian age, Auden, J.B. (1933) stated from Algonkian to Devonian, Rode, K.P. (1946) indicated early Palaeozoic age for the Rohtas limestone, Mishra, R.C. (1949) suggested Devonian age, Sitholey, R.V. (1953) assigned Cambrian age to the upper Vindhyan beds and Bose, A. (1956) found upper Cambrian age for the Kaimur and later Cambrian age for the Rewa. Vinogradov, A. and Tugarinov, A.J. (1964) have estimated the age of Glauconite sandstone as 1110 ± 60 million years and for Semri as between 1100 and 1400 million years. Salujha, S.K. (1982) dated the Kaimur between 910 and 940 million years. Srivastava, V.K. (1971) assigned Ordovician to Silurian age to the lower Vindhyan while Pichamuthu, C.S. (1971) has indicated that the base of the Vindhyan is probably 1400 million years. Salujha, S.K. (1982) concluded that Vindhyan sequence ranges from late pre-Cambrian to early Silurian. The lower Vindhyan is assigned in age from late pre-Cambrian to middle Cambrian and the upper Vindhyan ranges in age from late Cambrian to early Silurian.

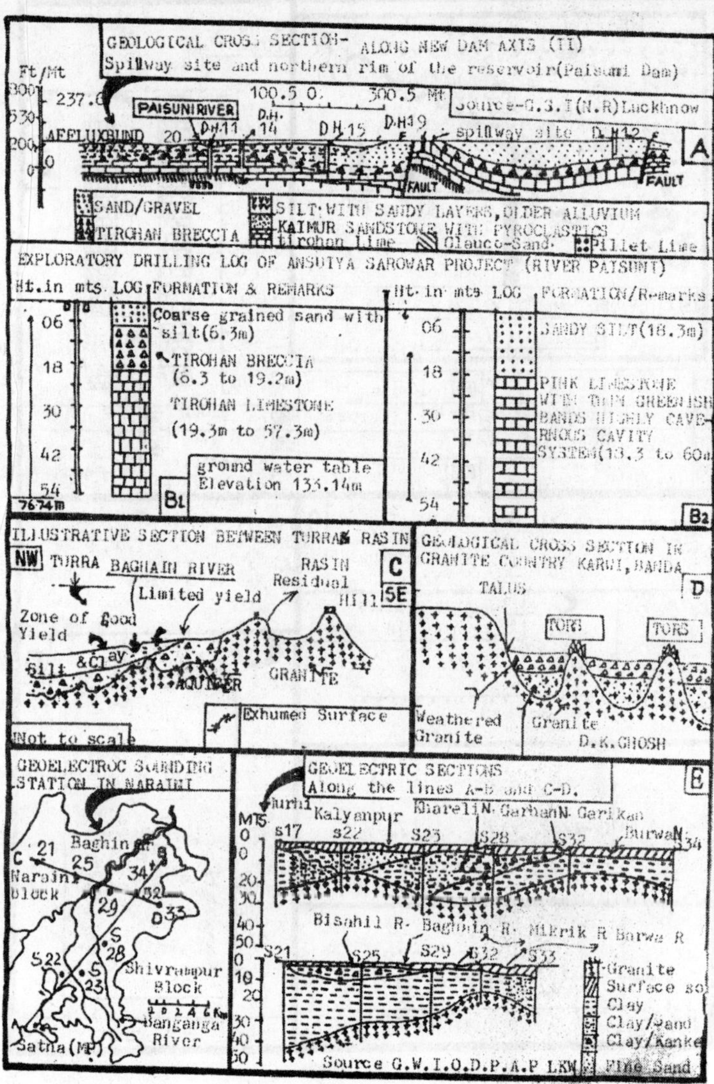
GEOLOGICAL CROSS SECTION- ALONG NEW DAM AXIS (II)
Spillway site and northern rim of the reservoir(Paisuni Dam)
Ft/Mt
237.6
PAISUNI RIVER
AFFLUX BUND
D.H.11
D.H. 14
D.H.15
D.H.19
spillway site
D.H.12
Source-G.S.I(N.R)Lucknow
FAULT
A
SAND/GRAVEL
TIROHAN BRECCIA
SILT WITH SANDY LAYERS,OLDER ALLUVIUM
KAIMUR SANDSTONE WITH PYROCLASTICS
tirohan Lime
Glauco-Sand
Pillet Lime
EXPLORATORY DRILLING LOG OF ANSUIYA SAROWAR PROJECT (RIVER PAISUNI)
Ht.in mts LOG FORMATION & REMARKS
Coarse grained sand with silt(6.3m)
TIROHAN BRECCIA (6.3 to 19.2m)
TIROHAN LIMESTONE (19.3m to 57.3m)
ground water table Elevation 133.14m
76.74m
B1
Ht. in mts LOG FORMATION/Remarks
SANDY SILT(18.3m)
PINK LIMESTONE WITH THIN GREENISH BANDS HIGHLY CAVERNOUS CAVITY SYSTEM(18.3 to 60m)
B2
ILLUSTRATIVE SECTION BETWEEN TURRA & RASIN
NW
TURRA
BAGHAIN RIVER
RASIN Residual Hill
Limited yield
Zone of Good Yield
Silt & Clay
AQUIFER
GRANITE
C
SE
Exhumed Surface
Not to scale
GEOLOGICAL CROSS SECTION IN GRANITE COUNTRY KARWI,BANDA
D
TALUS
TORS
TORS
Weathered Granite
Granite
D.K.GHOSH
GEOELECTROC SOUNDING STATION IN NARAINI
Baghin R
C
Naraini block
D
Shivrampur Block
Banganga River
Satna (MP)
GEOELECTRIC SECTIONS Along the lines A-B and C-D.
E
Kalyanpur
Khareli
N. Garhan
N. Gerikan
Burwal
Bisahil R.
Baghain R.
Mikrik R
Barwa R
Granite
Surface soil
Clay
Clay/Sand
Clay/Kanker
Fine Sand
Source G.W.I.O.D.P.A.P LKW

Fig. 1.13

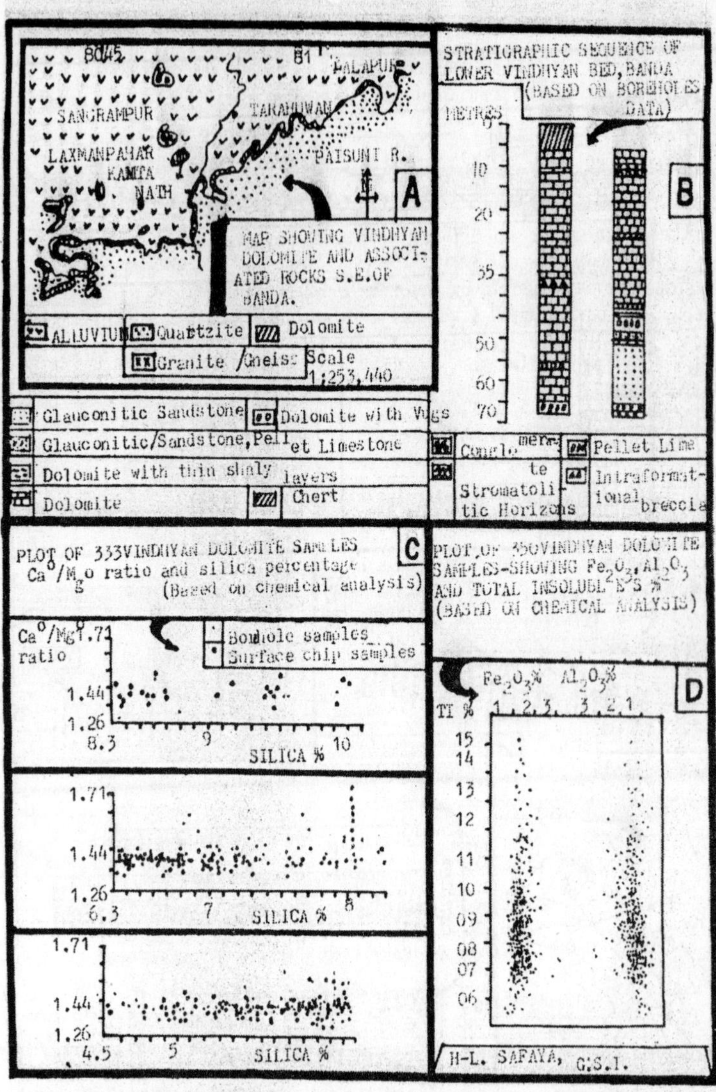
SANGRAMPUR
LAXMANPAHAR
KAMTA NATH
TARAHUWAN
PALAPUR
PAISUNI R.
A
MAP SHOWING VINDHYAN DOLOMITE AND ASSOCIATED ROCKS S.E. OF BANDA.
ALLUVIUM
Quartzite
Dolomite
Granite /Gneiss
Scale 1:253,440
STRATIGRAPHIC SEQUENCE OF LOWER VINDHYAN BED, BANDA (BASED ON BOREHOLES DATA)
METRES
B
Glauconitic Sandstone
Dolomite with Vugs
Glauconitic/Sandstone, Pellet Limestone
Dolomite with thin shaly layers
Dolomite
Chert
Conglomerate
Stromatolitic Horizons
Pellet Lime
Intraformational breccia
PLOT OF 333 VINDHYAN DOLOMITE SAMPLES Ca0/Mg0 ratio and silica percentage (Based on chemical analysis)
C
Ca0/Mg0 ratio
Borehole samples
Surface chip samples
SILICA %
PLOT OF 350 VINDHYAN DOLOMITE SAMPLES SHOWING Fe2O3, Al2O3 AND TOTAL INSOLUBLES %
(BASED ON CHEMICAL ANALYSIS)
D
Fe2O3% Al2O3%
TI %
H-L. SAFAYA, G.S.I.

Fig. 1.14

Bundelkhand complex as a topographic barrier in the heartland was exposed in pre-Cambrian age (2500 m.y. old to 2600 m.y. old) and form basement for the Vindhyan supergroup (1400 to 900 m.y. old). The Bundelkhand massif was the oldest platform of landform development. pre-Cambrian granitisation developed the pre-Vindhyan erosion surfaces in the region with mature landforms caused as a result of prolonged weathering where uplift of the granitised blocks, deformation and erosion were most common (Ravi Prakash and Dalela, I.K. 1982). Upon this surface the upper Vindhyan rocks were laid later. Prakash, R., Sharma, D.P. Srivastava, R.N. and Bhatt, G.D. (unpublished report D.G.M. 1972) stated that while a mature flat erosion surface developed in a prolonged cycle in pre-Vindhyan times, the cycle was shorter in duration and was interrupted by vertical movements along east-west trends, and the evaluation of isolated locally negative basins in which the oldest rocks of the Semri group were formed. Ghosh, D.B. (1976) stated that mature sandstones form a dominant phase with fresh activation and development of a larger basin over the mature erosion surface.

In the present study region the Tirohan group overlies an erosion surface of the Bundelkhand granite gniesses and is unconformably underlain by the Kaimur sandstone (Prakash, R. and Delela, I.K. 1982). Thus, the second cycle of landform development is cleared over Tirohan group. Recent work undertaken by the State Directorate of Geology on agates and clay deposits which form large pockets along a Tirohan erosion surface and below the Kaimur sandstone suggests the presence of a major break.

Prakash and Delela suggested that there was a linear control in the whole system. He stated that Semri basin was indeed confined and quite different from the later cyclic platform in which the Kaimur and Rewa were formed as transgressional facies of cover rocks. The Rewa group recently underlies by a new cyclic platform of Alluvium cover. Thus, it may be stated that the region experiences five platforms of different sediments like recent alluvium, Rewa group, Kaimur group, Tirohan group and basement granite from top to bottom. The upper Vindhyan surface is a dominantly arenaceous facies transgressively overlying the lower Vindhyan. A vast erosional unconformity between Kainur and Tirohan boundary in the area clearly indicates the two different cycle of landform development.

The top section of Kaimur near Manikpur along with the presence of volcanic breccia and Tuffaceous rocks suggests acid volcanism at the end of the Kaimur sedimentation (Prakash, R. and Dalela, I.K. 1982). The Vindhyan basin sediments in central and eastern part form a large gentle synclise and appear as successively arranged parabolic bands with closure at the north-eastern end. The significant feature of the Vindhyan outcrops is their sheet like geometry and conformity with the present margin of the basin. The two subdivisions of Vindhyan supergroup (i.e. lower and upper Vindhyan) mark a significant variation in their sequence and therefore represents that the beginning of Kaimur sedimentation as changes from predominantly carbonate to clastic sedimentation verifies a reflection of tectonic activity preceding the upper Vindhyan sedimentation (Chanda, S.K. and Bhattacharya, A. 1982). The lower Vindhyan sequence also witnessed relatively greater tectonic disturbance and magmatic activity as compared to the upper Vindhyan sequence. The Semri group is also known to be overlapped by the upper Vindhyan sediments. These relationships suggest existence of positive areas all around the eastern part of the basin excepting west where much remains concealed under the Decan trap (Chanda, S.K. and Bhattacharya, A. 1982). Banerjee and Sengupta postulated two sub-basins in the Kaimur times—a more actively subsiding euxinic eastern basin and a less actively subsiding open circulation, normal shallow marine western basin. Thus it appears that the basement ridge again became rejuvenated during the Kaimur time.

An exception (reversal of the regional north-westerly paleocurrent) has been reported by Banerjee, I. (1964) from basal sandstone and conglomerate of the Vindhyan supergroup occurring along the margin of the Bundelkhand massif, which is of particular significance with regard to Vindhyan sedimentation for it would suggest that the present northern limit of outcrops actually defines the margin of the basin and eliminates the possibility of the extension of the Vindhyan sediments below the Ganga basin at least during the lower Vindhyan period. Jafar, S.A., Akhtar, K. and Srivastava, V.K. (1966) also maintain that the lower Vindhyan sea did not extend much beyond the present northern limit of the Vindhyan outcrop; subsequent resumption of a regionally persistent north-westerly flow, of course, points to a possibility of northward transgression beyond the Bundelkhand massif in upper Vindhyan time. Such transgression is apparently supported by overlapping of the Semri group by the upper Vindhyan in the present study region and also by the westerly paleocurrent in Rajasthan just west of Bundelkhand massif. Banerjee, I. (1964) stated that the depocentres of the Vindhyan basin in the eastern part shifted towards north with time while Jafar, et al. (1966) postulated northward regression of the southern shoreline during upper Vindhyan time. The sediments below the unconformity in the Ganga valley are observed to be closely related to the upper Vindhyan specially to their upper part. This fact also indicates continuation of the Vindhyan towards north below the Ganga valley alluvium (Salujha, S.K. 1982). Hari Narayan and Kalia, K.L. (1982) stated that the Vindhyan sediments support the hypothesis of active Horst and graven tectonics during the Vindhyan sedimentation. During the lower Vindhyan times, irregularities in the basin floor were present in the form of basement ridges. These ridges were oriented both parallel to the basin margin and at right angles to it. A phase of volcanism during the lower Vindhyan period produced a sequence of pyroclastic sediments (rhyolitic tuffs, agglomerates and breccias). The phase of volcanism following a period of sedimentation in a stable shelf environment indicates a period of instability and some of the basement ridges below the lower Vindhyan might have been produced or accentuated in this period due to differential subsidence within the basin (Banerjee, I. 1964). Sharma, R.P. (1982) revealed that the Bundelkhand complex had undergone several cycles of deformations and metamorphism and the structure developed is a result of lateral compression and vertical movement of the creation. The style and attitude of folds, trends of foliation and domal features, north and north-east trending lineaments, west and north-west trending lineaments, faulted contact and regional tectonic framework correspond with the major tectonic of the Narmada-Son lineament, boundary faults of the Aravalli and normal faults of the Indo-Gangetic trough.

Geologists efforts regarding the tidal range, tidal circulation pattern, bedforms and depth of water in Vindhyan tidal sea are remarkable. Available evidence indicates that the entire Vindhyan sea was controlled by microtidal (tidal range 2 m.) occasionally rising to the mesotidal level (4-2 m.). On the open shelf towards the north-east, a shore normal and north-south biopolar tidal pattern were operated while within the embayed portion of the sea an amphidromic tidal system produced a complex tidal current pattern. Further complications were added by storm-surges and wind and wave generated currents. A significant feature of Vindhyan sequence is less likely in a microtidal regime and suggests a mesotidal range. Perhaps this indicates fluctuations in the tidal range during Vindhyan sedimentation (Banerjee, I. 1982). Analysis of paleocurrent data over the Vindhyan basin has shown the dominance of a unimodal north-westerly pattern (Banerjee, I.B. and Sengupta, S. 1963, Banerjee, I. 1964, Mishra, R.C. 1969, Banerjee, I.B. 1974).

In the Vindhyan barrier coastline, the bedform environments are well represented by the tidal flat, the lagoon and the barrier island. The washover flat has been also recognised (Banerjee, I.B. 1974). The depth of water in Vindhyan sea hardly exceeded 10 m. At the last phase of Vindhyan sedimentation (Bhander times) when all the lagoons behind three rows of barriers were filled up and the entire basin was converted into an extensive tidal flat, the depth of water must have been a few metres only (Banerjee, I. 1982).

On the basis of the above discussion, it may be stated that the Vindhyan were deposited on the peneplained. Archaen III (2500-3200 m.y.) and lower Proterozoic (1800-1200 m.y. to over 2500 m.y., older than Delhi and Satpura folding) basement rocks (pre-Aravalli metasediments) having irregular topography. At that time Vindhyan mountains in form of gneissic ridge and the Aravalli range had already taken shape (Prasad, B. 1984). As an aftermath of the Eastern Ghat orogeny (1600 ± 1000 m.y.), platform conditions developed around the Bundelkhand shield and the southern depression became a site of prolonged shallow intracontinental marine sedimentation (1400-600 m.y.) represented by the Vindhyan. A tectonic event of epeirogenic nature not only terminated the process of sedimentary accumulation, but gave rise to long deep normal faults represented by the Son-Narmada faults and the Great Boundary Fault of Rajasthan. Under the northerly sloping cration, hot spots developed towards the close of the palaeozoic giving rise to a triple junction, characterized by all aborted rifts down which flowed rivers in the north-westerly and westerly directions and deposited thick continental Gondawana sediments. At the end of the mesozoic and beginning of cenozoic there was a great outpouring of lavas giving rise to the Deccan trap. Following the Himalayan orogeny, the northern part of the platform sagged phenomenally to give rise to enormous thickness of the Ganga Basin (Valdiya, K.S. 1982).

Thus, the entire platformal Vindhyan sedimentation took place under an epeirogenic condition caused by change in the isostatic balance on account of rejuvenation of Aravalli range. As is evident from above account, it appears that the rejuvenation of Aravalli range during post to Delhi sedimentation is primarily responsible for the platformal deposition of the Vindhyan and which came to a halt with the cessation of rise of the Aravalli range following the extrusion and intrusion of the Malani suite of Igneous rocks. Due to lack of relief the sedimentation reached the closing phase and a continental condition prevailed. Over 3200 m. thick Vindhyan sediments were deposited in over 700 to 800 m.y. which indicates a very slow rate of sedimentation, at the rate of only 4.5 to 4.0 m. per million year. This suggests a limited area of erosion, poor weathering agencies and low relief of provenance vis-a-vis site of deposition. This in turn is indicative of palaeogeographic condition of the Vindhyan sedimentation which took place mainly under oscillatory basin condition (Prasad, B. 1984).

Apart from the above, Singh, S.N. and Pal, O.P. (1970) have illustrated some classical views of Chitrakut region on the basis of detailed field study of the region. They suggested that at least during the lower Vindhyan deposition, the Chitrakut basin seems to be a small separate part of the main epeirogenic Vindhyan basin. The lower Vindhyan sea for the first time inundated this area, when perhaps the olive shales or the Glauconitic sandstones were being deposited in the Son valley region. The development of Kamtanath member type litho-unit at different heights in various hills in the northern part seem to be the result of a gradually transgressing sea, the environment being a shelf area with moderately strong wave action perhaps produced partially by tide action and the other members of the Chitrakut formation were being simultaneously deposited in the deeper part of the basin.

The Semri deposition started at the margins of the shelves on an undulating granite topography. The transgressional phase from east and south-east continued with the result that in the northern part of the area, the Kamtanath type of litho-unit was formed at higher levels fringing the hill slopes. In this series of gradual transgression phase, there was a comparatively stronger phase inundating the area suddenly with the development of glauconitic orthoquartzite at the top of the Lodhwara hill directly overlying the granite. The glauconitic orthoquartzite of Lodhwara hill probably indicates a period of minor regression. The major part of the Tirohan limestone member was perhaps formed during the succeeding transgressive phase after this minor regression. The end of Tirohan limestone deposition marks a complete regression on the sea from this part of the basin as represented by disconformity and further the area was inundated by stronger Kaimur transgression with the result that now the Kaimur sandstone is seen overlap all pre-Kaimur lithological types of this area.

Drainage System

Chitrakut region represents an unique type of drainage system which constitutes the most important element of surface geodynamics. Paisuni, a major tributary of Yamuna river constitutes the major drainage system of the region. It covers more than 80 per cent of the total surface area of the region. Ohan, Banganga and Ganta are also the major tributaries which form the system. The nature and characteristics of major drainage lines of the region are given in Fig. 1.15 and Tables 1.2 and 1.3.

The various drainage patterns of the region are shown in Fig. 1.15. The most commonly encountered drainage pattern is dendritic which is characterised by irregular branching of the tributary streams in many directions and almost at any angle although usually at considerably less than a right angle. Following to Fig. 1.15, it may be explained that the drainage basins like Samrani, Paroi, Gauhia, Kachchua, Dhora, Ganta, Kewai, Kulagara, Barar, Pindrah, Kalibarah, Dingchi and Ghilhal are of dendritic pattern (D). Drainage basins like Sukra, Maro, Bagdara and Jhuri register approximately a rectangular pattern. Contorted pattern can be observed in the lower reaches of Ghinhal nala. Barwa and the major right bank tributaries of Banganga are of 'Radial' pattern. Parallel patterns are marked along the small tributaries of Banganga which generally run parallel in nature. All these patterns are principally governed by topographic and lithological characteristics of the country concerned (Prasad, G. 1987).

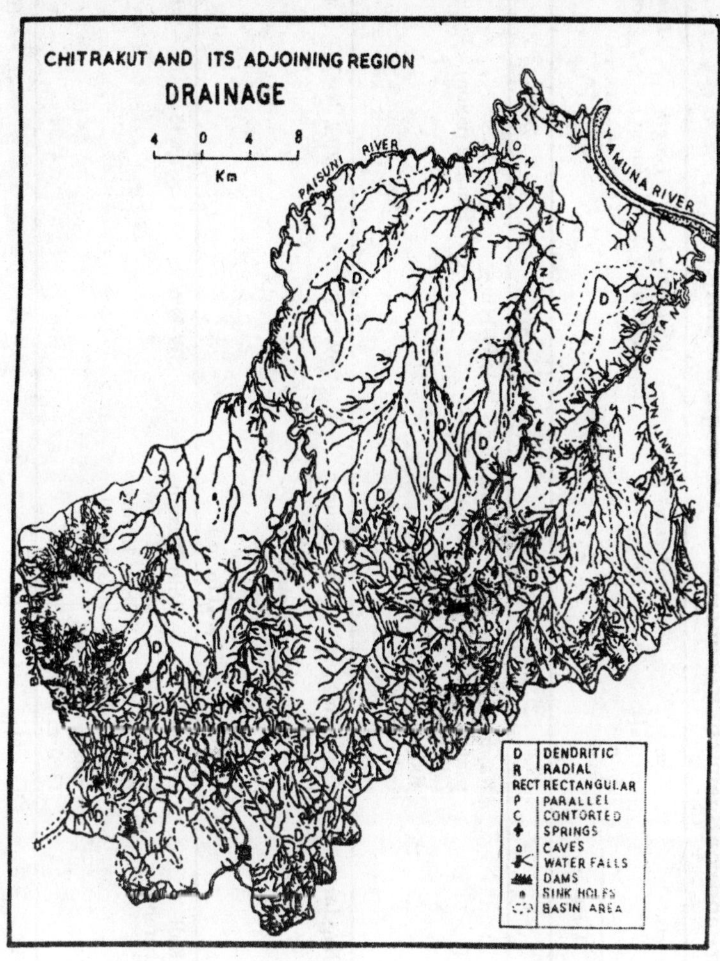
CHITRAKUT AND ITS ADJOINING REGION
DRAINAGE
4 0 4 8
Km
PAISUNI RIVER
YAMUNA RIVER
D DENDRITIC
R RADIAL
RECT RECTANGULAR
P PARALLEL
C CONTORTED
SPRINGS
CAVES
WATER FALLS
DAMS
SINK HOLES
BASIN AREA

Fig. 1.15

Table 1.2: Source and confluence of Paisuni river and its major tributaries

S. No.	*Streams*		*Source (Height in m.)*	*Concluence (Height in m.)*	*Flow direction*	*Length (in m.)*	*Drainage order*
	Major	*Tributary*					
1	*2*	*3*	*4*	*5*	*6*	*7*	*8*
	Paisuni		Kaimur plateau (above 400 m., out of the region)	With Yamuna near Kankota village (80 m.)	SW-NE	136.95 (in the region)	–
1.		Bagdara	Chitrakut plateau (350 m.)	With Paisuni about 1 km. south of Kothria village (176 m.)	SW-NE	20.2	4th
2.		Kali-barah	South of Chitrakut plateau near Gogla village (360 m.)	With Paisuni about 1 km. SE of Kothria village (180 m.)	SW-NW	13.55	5th

(Contd...)

1	2	3	4	5	6	7	8
3.		Maro	South-eastern part of the plateau near Manikpur (316 m.)	With Paisuni about 1 km. south of Sati Anusuiya hill (160 m.)	SE-NW		
4.		Jhuri	South-eastern part of the plateau near Bharhlyan village (340 m.)	With Paisuni near Darwaza village (152 m.)	SW-NE	17.50	4th
5.		Samrani	West of Bhitakhera village in north plain (120 m.)	With Paisuni near Lohda village (93 m.)	SW-NE	29.50	3rd

(Contd...)

1	2	3	4	5	6	7	8
6.		Ohan	South of the plateau near Markundi (277 m.)	With Paisuni near Sagwara village (95 m.)	SE-NW	72.30	–
1.	Ohan	Sukra	South-eastern part of the plateau about 1 km. north of Manikpur (270 m.)	With Ohan near Pandit ka Parwa (190 m.)	S-N	–	–
2.		Kulagara	South-eastern part of the plateau near Barahmafi village (260 m.)	With Ohan near Bahil-Purwa village (175 m.)	SE-NW	11.40	4th

(Contd...)

1	2	3	4	5	6	7	8
3.		Pindrah	Mandali Dhawar hill (306 m.)	With Ohan near Bauna village (155 m.)	SW-NE	11.80	3rd
4.		Barar	Jati Pahar (268 m.)	With Ohan near Char village (126 m.)	S-N	17.90	5th
5.		Gauhia	Hillock country near Bishawan village (140 m.)	With Ohan near Byas Barna village (94 m.)	SW-NE	13.30	3rd
6.		Geduwa	Kottirth hill (305 m.)	With Ohan near Kalwalia village (95 m.)	SW-NE	41.30	–
	Geduwa						
1.		Piprawal	Kottirth hill (305 m.)	With Geduwa near about 1.5 NW of Ragauli village (116 m.)	S-N	17.20	3rd

(Contd...)

1	2	3	4	5	6	7	8
2.		Dhora	North of Matdar hill (297 m.)	With Geduwa about 1.5 NW of Ragauli village (117 m.)	S-N	15.10	3rd
3.		Paroi	Baruhai Pahar (160 m.)	With Geduwa near Bakta village (96 m.)	S-N	27.45	4th

Table 1.3: Source and confluence of Ganta and its major tributaries

S. No.	*Streams*		*Source (Height in m.)*	*Confluence (Height in m.)*	*Flow direction*	*Length (in m.)*	*Drainage order*
	Major	*Tributary*					
1	2	3	4	5	6	7	8
	Ganta		Manikpur hills from Tiwaribaba Pahar (277 m.)	With Yamuna about 1 km. south of Tirmau village (83 m.)	S-N	51.7	–
1.		Ganta	Manikpur hills from Tiwaribaba Pahar (277 m.)	With Ganta major channel near Garhchapa village (140 m.)	SW-NE	4.8	4th
2.		Kewai	Khan Banha hill near Manikpur (260 m.)	With Ganta near Raipura village (140 m.)	SW-NE	22.75	5th

(Contd...)

1	2	3	4	5	6	7	8
3.		Jhurai	South-eastern part of the plateau near Harhai village (235 m.)	With Ganta near Kapuri village (140 m.)	SE-NW	19.80	3rd
4.		Jaiwanti	Twan hills in the south-eastern part of the plateau (231 m.)	With Ganta near Amarpur village (90 m.)	S-N	–	3rd
5.		Kachchua	Northern plain near Raipura village (115 m.)	With Ganta near Atarsui village (83 m.)	SW-NE	18.90	3rd

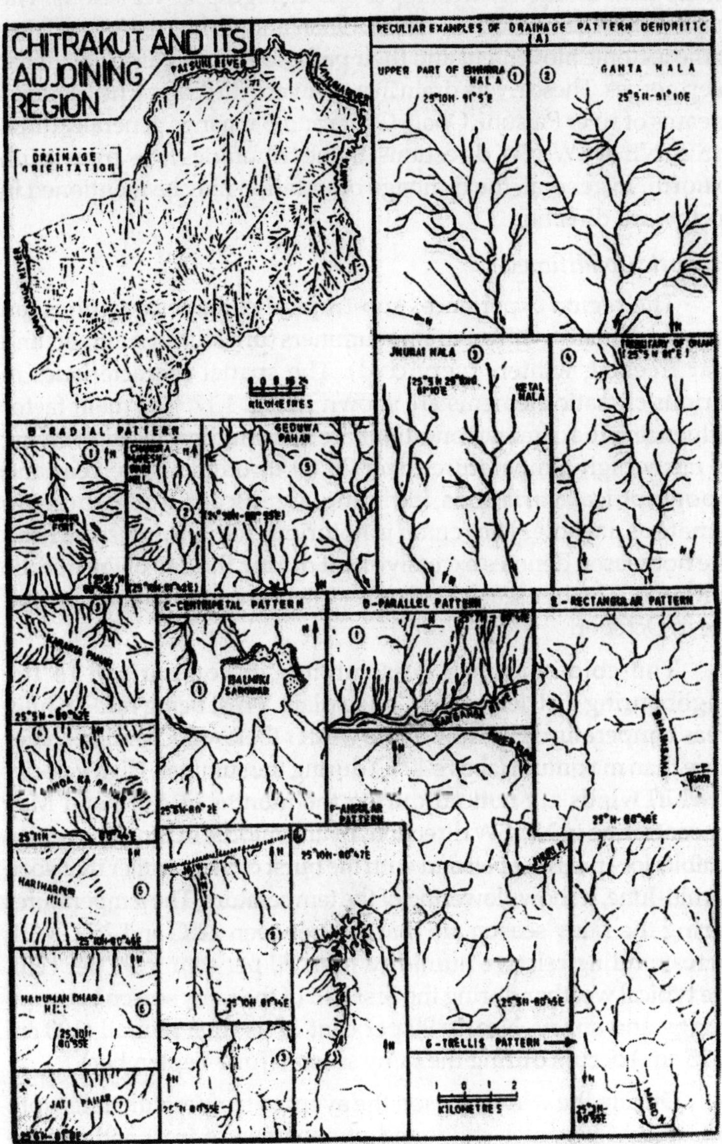
CHITRAKUT AND ITS ADJOINING REGION
DRAINAGE ORIENTATION
KILOMETRES
PECULIAR EXAMPLES OF DRAINAGE PATTERN DENDRITIC (A)
UPPER PART OF BHIRRA NALA
GANTA NALA
JHURAI NALA
B-RADIAL PATTERN
GEDUWA PAHAR
C-CENTRIPETAL PATTERN
D-PARALLEL PATTERN
E-RECTANGULAR PATTERN
F-ANGULAR PATTERN
HARIHARPUR HILL
HANUMAN DHARA HILL
JATI PAHAR
G-TRELLIS PATTERN
KILOMETRES

Fig. 1.16

The drainage orientations of the region are influenced by lithological characteristics slopes and topographic variations. The major drainage lines like Paisuni, Ohan and Ganta are formed due to the tectonic movement and their paths are much guided by linear deep gorges. These rivers drain in northward direction. The tributary streams of river Paisuni, Ohan, Ganta and Banganga generally drain in SW-NE or SW-NW directions. Some tributaries flow from south to north. In general, the drainage orientation may be mentioned in northward direction.

Climatic Conditions

The region experiences sub-tropical climatic conditions (an extreme climate) with a extreme summers (prickly hot summer) and near freezing winters (very cold). The spatial characteristics in various climatic elements are shown in Fig. 1.17. The main factor influencing the climatic conditions are its latitudinal location giving its fairly high temperature (over 25°C) throughout the year and topographical variations initiating significant fluctuations in climate as a whole. The region is notorious for its oppressive heat. The hot season denotes excessive heat during the day of the middle of March. The hot season becomes intense very soon and last till late in October.

The cold season is less intense in comparison to the neighbouring districts of the Ganga plain. Frost being rare. In hilly areas temperature goes low in the winter (below 5°C) and becomes more than maximum (above 40°C) during the summer. High velocity of whirl winds are noticed during the months of April and May concentrating in NE-SW direction. By the middle of June, the region exhibits low pressure belt and with the burst of the summer monsoon by mid-June, it shows lowering of the temperature. The temperatures during the rainy season are marked between 22°C and 25°C with corresponding relative humidity from 70 per cent to 90 per cent. The typical weather during the first half of the rainy season becomes muggy and sultry. Nearly 90 per cent of its total rainfall (100 cm - 125 cm.) occurs during the rainy season (June-September).

During the winter season the average temperature falls upto 15°C. Night becomes chilly and phenomena of frost and fog are common. The rains due to westerly disturbance during the winter season are highly beneficial for the Rabi crops.

The period of the monsoon is divided into the period of general rains from mid-June to mid-September and the period of retreating monsoon from mid-September to December. On the basis of the above characteristics, a year is divided into the three main seasons:

1. The hot weather season (March to mid-June).
2. The rainy season (Mid-June to mid-October).
3. The cold weather season (Mid-October to December).

Soils

The soils of 'Chitrakut Upland' are principally governed by various environmental factors like parent material, time, topography, climate, vegetation and living organisms. These factors have changed mainly the composition and fertility of the soil in a number of ways from one place to other (Prasad, G. 1986).

On the basis of extensive field study and consulting several agricultural officials, the soils of the region have been grouped into the following categories (Fig. 1.18):

1. Upland soil –	(i)	Gravelly soil (Rankar)
2. Low land soil –	(ii)	Coarse grained brown soil (Paura)
	(iii)	Shallow black soil (Light Mar or Kabar)
3. Riverine soil –	(iv)	Tari
(Alluvial)	(v)	Kachchar; and
	(vi)	Rankar

Gravelly soil occupies about 2/3 land of Chitrakut plateau specially over the geological formations of Tirohan limestone and Kaimur sandstone. Due to presence of maximum proportions of gravel it is named as gravelly and locally termed as 'Patha' soil. It includes the poorer form of 'Paura', Mar and Kabar with texture varying from clay loam to sandy loam. It contains high pH(8), little humps, well leaching capacity and high degree of porosity. The soil becomes exhausted when cultivated hence when continuously cropped fertilization becomes necessary. These soils are highly

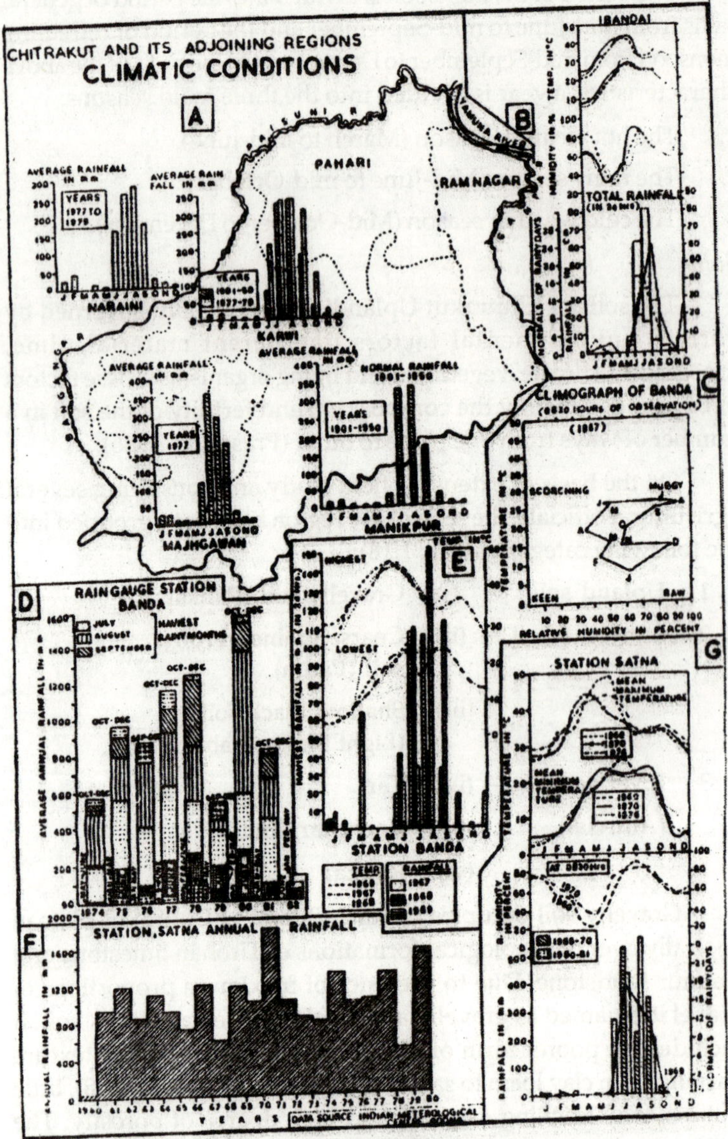
CHITRAKUT AND ITS ADJOINING REGIONS
CLIMATIC CONDITIONS
A
B
C
D
E
F
G
PAHARI
RAMNAGAR
YAMUNA RIVER
NARAINI
KARWI
MAJHGAWAN
MANIKPUR
BANDA
CLIMOGRAPH OF BANDA
RAINGAUGE STATION BANDA
STATION BANDA
STATION SATNA
STATION, SATNA ANNUAL RAINFALL
DATA SOURCE INDIAN METEROLOGICAL

Fig. 1.17

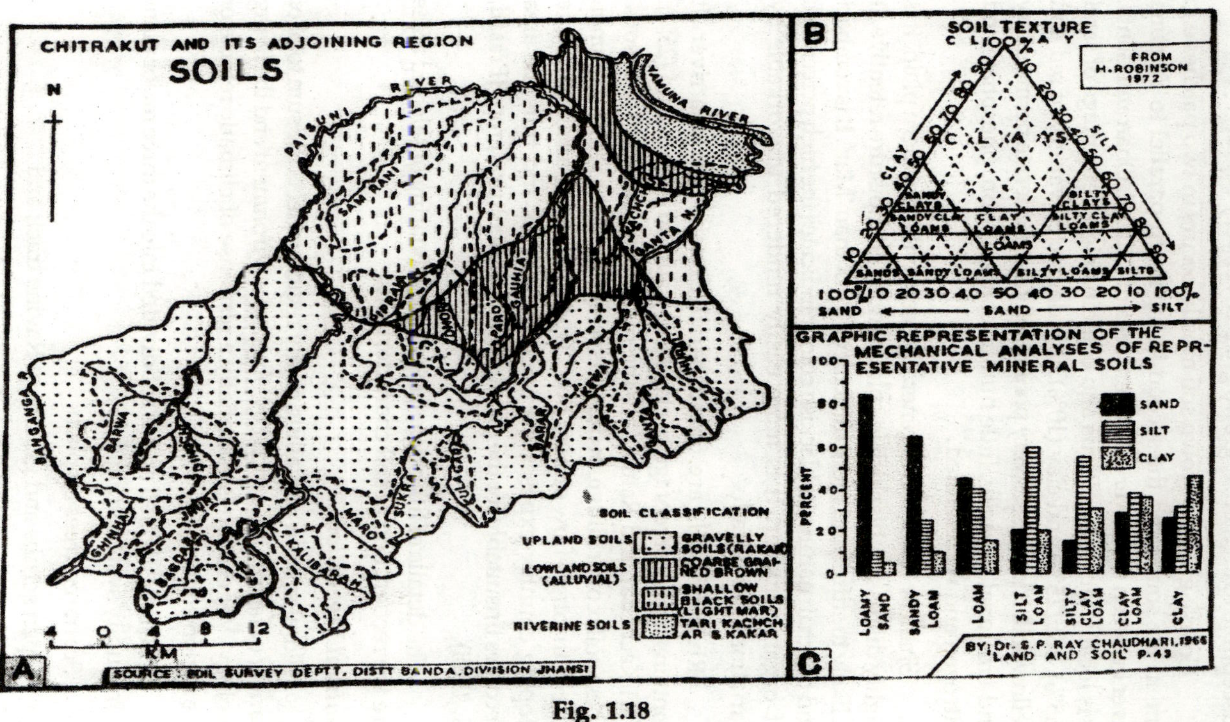
A
CHITRAKUT AND ITS ADJOINING REGION
SOILS
N
YAMUNA RIVER
4 0 4 8 12
KM
SOIL CLASSIFICATION
UPLAND SOILS
GRAVELLY SOILS (RAKAR)
LOWLAND SOILS (ALLUVIAL)
COARSE GRAINED BROWN
SHALLOW BLACK SOILS (LIGHT MAR)
RIVERINE SOILS
TARI KACHCHAR & KAKAR
SOURCE : SOIL SURVEY DEPTT. DISTT BANDA, DIVISION JHANSI
B
SOIL TEXTURE
FROM H. ROBINSON 1972
CLAYS
SANDY CLAYS
SILTY CLAYS
SANDY CLAY LOAMS
CLAY LOAMS
SILTY CLAY LOAMS
LOAMS
SANDS
SANDY LOAMS
SILTY LOAMS
SILTS
SAND
SILT
C
GRAPHIC REPRESENTATION OF THE MECHANICAL ANALYSES OF REPRESENTATIVE MINERAL SOILS
SAND
SILT
CLAY
PERCENT
LOAMY SAND
SANDY LOAM
LOAM
SILT LOAM
SILTY CLAY LOAM
CLAY LOAM
CLAY
BY: Dr. S.P. RAY CHAUDHARI, 1966 LAND AND SOIL P.43

Fig. 1.18

susceptible to erosion. It grows grass cover very rapidly.

Coarse grained brown soil mostly occurs in two patches e.g. one in the south and the second in the north parallel to Yamuna river between the lower catchment area of Paisuni, Ohan and Ganta. It is locally termed as Parua and contains high percentage of sand (65%), silt (25%) and clay (10%). It is degraded variety of red and yellow soil. It possesses very poor reactivity, moderate permeability and erosional capacity, little humus content, iron, phosphate and nitrogen and well aerated.

Shallow black soil is generally seen in the northern plain. It is highly disfussed, blackish in colour and high moisture retentive. It is likely to stiffer and more difficult to work than 'Mar'. It is compact and tough in nature. Its blackish colour indicates the high percentage of organic matter. It is more easily rendered unworkable by variations in rainfall.

Riverine soils are found parallel to the Yamuna river. It contains high percentage of clay (45%), silt (30%) and sand (25%). Tari Kachchar and Ranker are the major categories of these soils. These soils cover alluvium deposits where gullying is most common due to through-flow or pipe flow. Due to gullying and erosion the sloping country exposes the calcium nodules and thus the area becomes unsuitable for the production of agricultural crops (Prasad, G. 1985).

The details of lithological logs of trial boaring of selected areas are the recognition of soil characteristics.

Natural Vegetation

Floral system prevailing in any region reflects the sum total of variety of factors influenced by physical environment and the living ecology. Vegetation is essentially a response of climatic regime of the region concerned.

The investigated area is subdivided into the three major plant community systems (Fig. 1.19):

1. Scattered trees on plain;
2. Open scrubs and grasses of Ravine tract; and
3. Dense mixed deciduous forest of Upland.

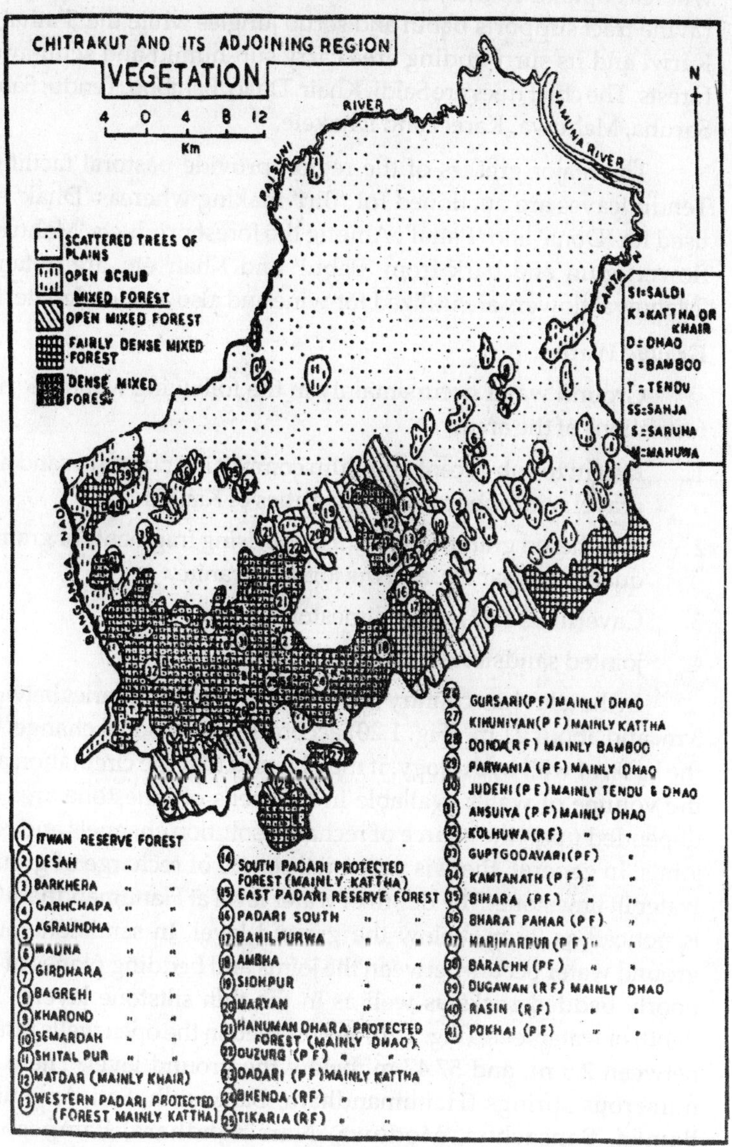
CHITRAKUT AND ITS ADJOINING REGION
VEGETATION
4 0 4 8 12
Km
N
RIVER
PAISUNI
YAMUNA RIVER
GANTA
BANGANGA NADI
SCATTERED TREES OF PLAINS
OPEN SCRUB
MIXED FOREST
OPEN MIXED FOREST
FAIRLY DENSE MIXED FOREST
DENSE MIXED FOREST
SI=SALDI
K=KATTHA OR KHAIR
D=DHAO
B=BAMBOO
T=TENDU
SS=SANJA
S=SARUNA
M=MAHUWA
1 ITWAN RESERVE FOREST
2 DESAH
3 BARKHERA
4 GARHCHAPA
5 AGRAUNDHA
6 MAUNA
7 GIRDHARA
8 BAGREHI
9 KHAROND
10 SEMARDAH
11 SHITAL PUR
12 MATDAR (MAINLY KHAIR)
13 WESTERN PADARI PROTECTED (FOREST MAINLY KATTHA)
14 SOUTH PADARI PROTECTED FOREST (MAINLY KATTHA)
15 EAST PADARI RESERVE FOREST
16 PADARI SOUTH
17 BAHILPURWA
18 AMBHA
19 SIDHPUR
20 MARYAN
21 HANUMAN DHARA PROTECTED FOREST (MAINLY DHAO)
22 OUZURG (P F)
23 DADARI (P F) MAINLY KATTHA
24 BHENDA (RF)
25 TIKARIA (RF)
26 GURSARI (P F) MAINLY DHAO
27 KIHUNIYAN (P F) MAINLY KATTHA
28 DONDA (R F) MAINLY BAMBOO
29 PARWANIA (P F) MAINLY DHAO
30 JUDEHI (P F) MAINLY TENDU & DHAO
31 ANSUIYA (P F) MAINLY DHAO
32 KOLHUWA (R F)
33 GUPTGODAVARI (P F)
34 KAMTANATH (P F)
35 BIHARA (P F)
36 BHARAT PAHAR (P F)
37 HARJHARPUR (P F)
38 BARACHH (P F)
39 DUGAWAN (R F) MAINLY DHAO
40 RASIN (R F)
41 POKHAI (P F)

Fig. 1.19

The northern alluvial plain is covered with scattered trees whereas upland country is dominated by dense mixed forest. The ravine tract supports Babul and scrub jungles while the 'Patha' of Karwi and its surrounding area carry sub-humid and semi-aried forests. The chief trees are Saldi, Khair, Dhao, Bamboo, Tendu, Sahja, Saruha, Mahuwa, Kareel and Dhakete.

The major grasses of the region provide pastoral facilities. Tendu leaves are auctioned for 'Biri' making whereas 'Dhak' are used for 'Dona' and 'Pattal'. Among the forest products 'Mahuwa' flower, gum and bark from 'Babul' and Khair etc. are notable. 'Mahuwa' flowers are utilized for wine and also as a food material.

Ground Water

Ground water occurs mainly in the following water bearing formations of the area:

1. Recent to sub-recent alluvium comprising clay, silt, sand and gravel with subordinate quantities of Kanker.
2. Weathered granitic residum comprising fragments of granite, quartz, felspar etc. and the jointed granite.
3. Cavernous and jointed limestone.
4. jointed sandstone (Kaimur sandstone).

In the alluvium country the depth of water level varies between 3 m. and about 21 m. (Fig. 1.20) according to seasonal change. On the basis of well hydrology, it may be said that the circulation and the volume of water available in the wells of limestone area are depended upon the source of recharge solution channels and rock joints. In general, there is perennial source of recharge of ground water in limestone. The depth of water level at Hanumandhara hill is noticed as 26 m. below the ground level. In sandstone area, ground water occurs between the joints and bedding planes of the poorly bedded rocks as well as in the thin siltstone layers. The depth of water level (Fig. 1.20) as observed in the open wells ranges between 2.5 m. and 57.47 m. below the ground level. There are numerous springs (Hanumandhara, Bankesidh, Sati Anusuiya, Bamka, Bansachua, Modhwa, Narsinghdhara, Pampeswar, Devangana and Kottirth) seepages and waterfalls as a ground water resource in the plateau region. These are the major source of water supply. The water level rises during the monsoon period and decreases in post-monsoon. The lowest level can be marked in the last weak of May and June (pre-monsoon period). The average fluctuations (Fig. 1.21 and 1.22) have been noticed to vary between 3 m.—5 m. in alluvium, 4 m. in shale and 6 m. in sandstone (Prasad, G. 1986). The maximum fluctuations (14.10 m.) are noticed in Majhgawan block (Satna district).

Monsoon recharge is measured as 35 per cent in alluvial areas, and 25 per cent in hard rocks area. Naraini, Karwi, Pahari, Ram Nagar and Manikpur blocks indicate 16.90 per cent, 11.56 per cent, 11.23 per cent, 38.99 per cent and 12.92 per cent of rainfall recharge by monsoon rainfall. Gross recharge (in ham) has been found as 14441 and 15595 in Naraini, 7342 and 9460 in Karwi, 7789 and 10213 for Pahari, 13189 and 6275 in Ram Nagar and 15205 and 17835 in Manikpur by fluctuation and rainfall infiltration approach. Net recoverable recharge has been taken as 70 per cent of gross recharge.

Ground water regimes are highly alkaline mainly in the alluvium and weathered granite mantle. Chorides are usually well within the drinking limit. The Bi-carbonate contents vary very widely and it is generally nil. Water is very hard. The ground water contours and well hydrographs (Fig. 1.22) are the true indicator of ground water balance. The suitable zone for ground water structures, its availability and utilization in different part of the study region are also shown in Figs. 1.21 and 1.22 respectively.

Cultural Ecosystem

The cultural ecosystem of Chitrakut region is still primitive in most parts. The people of 'Patha' area and the remote parts still eat wild fruits and berries, go for hunting and collecting forest products. They rely heavily on nature for their many existence. Man does not concern himself with conservation methods. Nature is able to cope with whatever damage is inflicted quickly restoring the balance.

The natural eco-system of the region has been greatly responsible in shaping the cultural eco-system. The Vindhyan scarps with numerous springs and the ravinous tracts with rivers

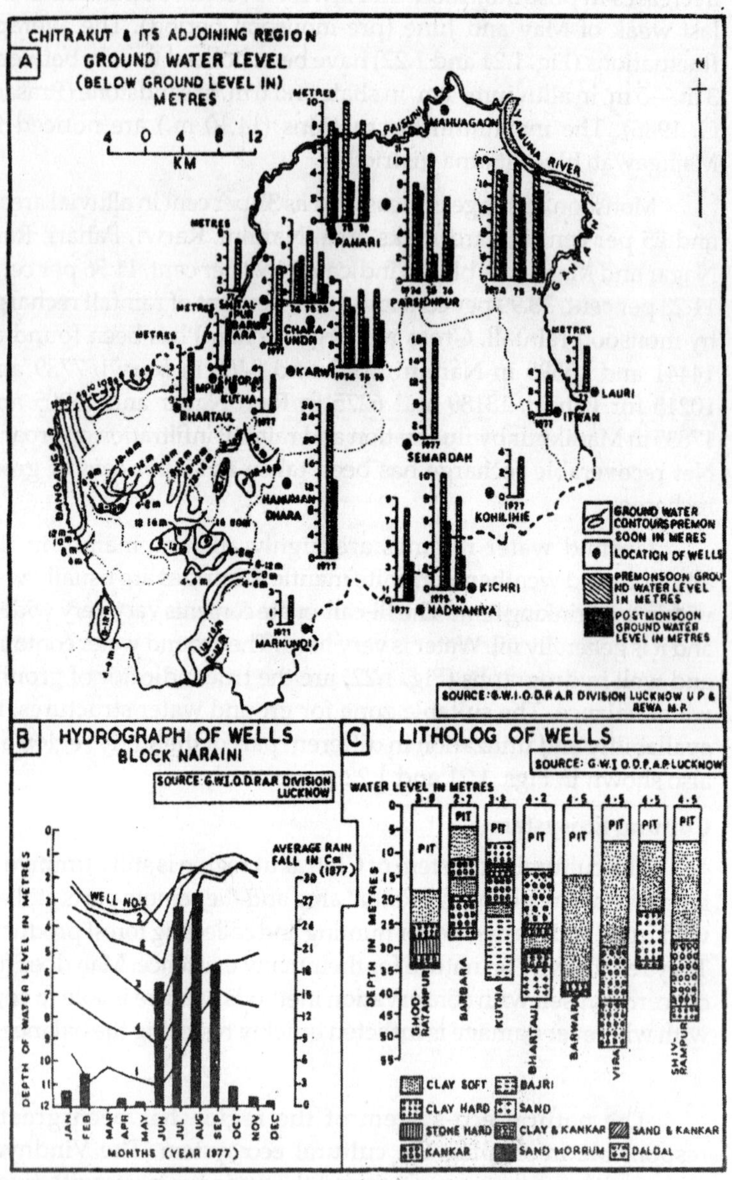
CHITRAKUT & ITS ADJOINING REGION
A
GROUND WATER LEVEL
(BELOW GROUND LEVEL IN)
METRES
KM
N
MAHUAGAON
PAISUNI R.
YAMUNA RIVER
PAHARI
PARSIDHPUR
BHARTAUL
KUTHA
CHARA UNDH
KARWI
SEMARDAH
LAURI
ITWAN
HANUMAN DHARA
KOHILINIE
KICHRI
NADWANIYA
MARKUNDI
BANGANGA
GROUND WATER CONTOURS PREMON SOON IN MERES
LOCATION OF WELLS
PREMONSOON GROUND WATER LEVEL IN METRES
POSTMONSOON GROUND WATER LEVEL IN METRES
SOURCE: G.W.I.O.D.R.A.R DIVISION LUCKNOW U P & REWA M.P.
B
HYDROGRAPH OF WELLS
BLOCK NARAINI
SOURCE G.W.I.O.D.R.A.R DIVISION LUCKNOW
AVERAGE RAIN FALL IN Cm (1977)
WELL NO.5
DEPTH OF WATER LEVEL IN METRES
JAN FEB MAR APR MAY JUN JUL AUG SEP OCT NOV DEC
MONTHS (YEAR 1977)
C
LITHOLOG OF WELLS
SOURCE: G.W.I O.D.P.A.P.LUCKNOW
WATER LEVEL IN METRES
DEPTH IN METRES
PIT
GHOD-RATANPUR
BARBARA
KUTNA
BHARTAUL
BARWARA
VIRA
SHIV-RAMPUR
CLAY SOFT
BAJRI
CLAY HARD
SAND
STONE HARD
CLAY & KANKAR
SAND & KANKAR
KANKAR
SAND MORRUM
DALOAL

Fig. 1.20

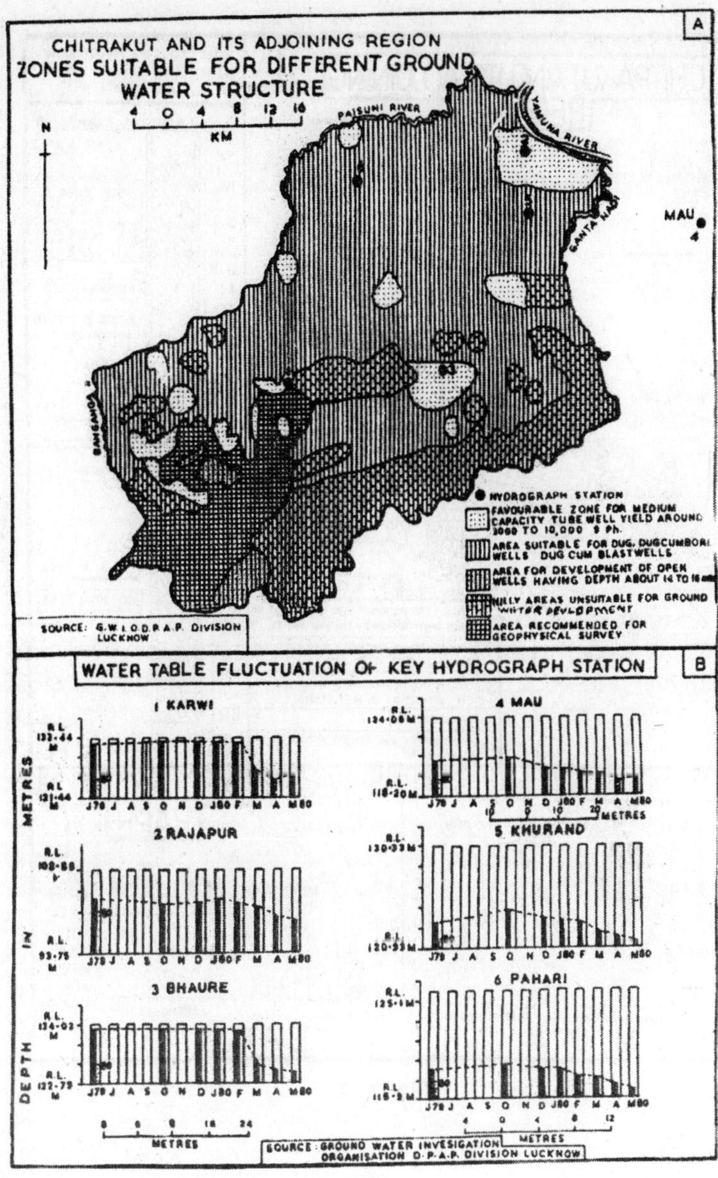
A
CHITRAKUT AND ITS ADJOINING REGION
ZONES SUITABLE FOR DIFFERENT GROUND WATER STRUCTURE
KM
PAISUNI RIVER
YAMUNA RIVER
MAU
SOURCE: G.W.I.O.D.P.A.P. DIVISION LUCKNOW
HYDROGRAPH STATION
FAVOURABLE ZONE FOR MEDIUM CAPACITY TUBE WELL YIELD AROUND 3000 TO 10,000 $ Ph.
AREA SUITABLE FOR DUG, DUGCUMBORI WELLS DUG CUM BLASTWELLS
AREA FOR DEVELOPMENT OF OPEN WELLS HAVING DEPTH ABOUT 14 TO 18
HILLY AREAS UNSUITABLE FOR GROUND WATER DEVELOPMENT
AREA RECOMMENDED FOR GEOPHYSICAL SURVEY
B
WATER TABLE FLUCTUATION OF KEY HYDROGRAPH STATION
1 KARWI
2 RAJAPUR
3 BHAURE
4 MAU
5 KHURAND
6 PAHARI
DEPTH IN METRES
METRES
SOURCE: GROUND WATER INVESIGATION ORGANISATION D.P.A.P. DIVISION LUCKNOW

Fig. 1.21

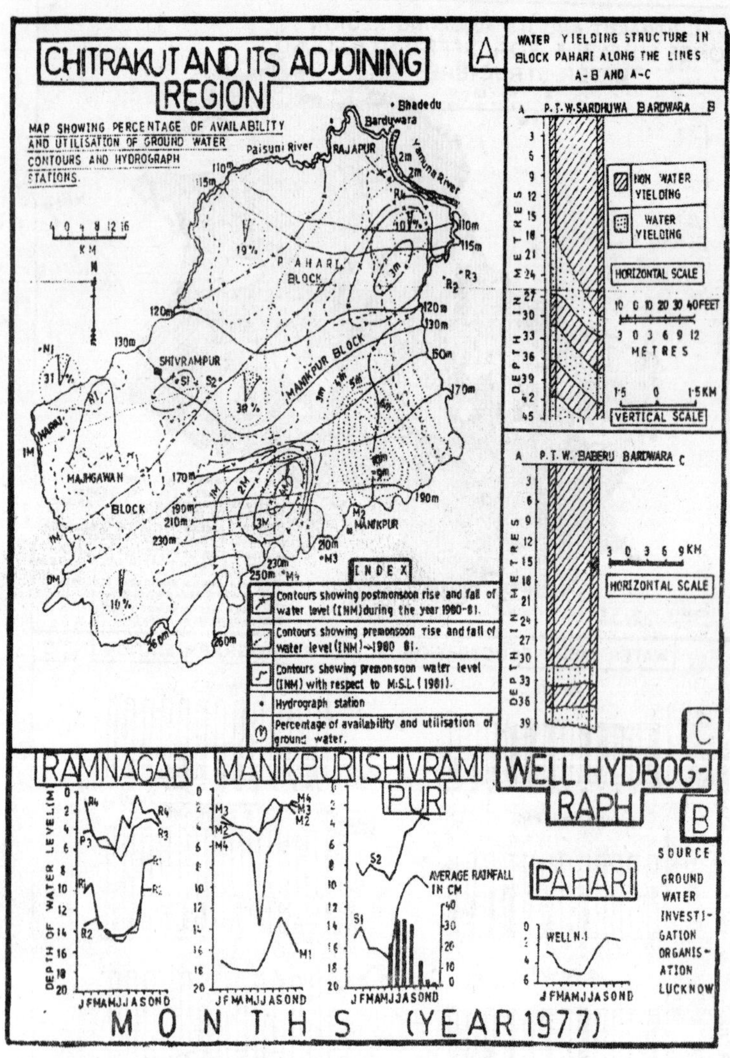
CHITRAKUT AND ITS ADJOINING REGION
MAP SHOWING PERCENTAGE OF AVAILABILITY AND UTILISATION OF GROUND WATER CONTOURS AND HYDROGRAPH STATIONS.
A
WATER YIELDING STRUCTURE IN BLOCK PAHARI ALONG THE LINES A-B AND A-C
P. T. W. SARDHUWA BARDWARA
NON WATER YIELDING
WATER YIELDING
HORIZONTAL SCALE
VERTICAL SCALE
DEPTH IN METRES
P. T. W. BABERU BARDWARA
PAHARI BLOCK
MANIKPUR BLOCK
MAJHGAWAN BLOCK
SHIVRAMPUR
RAJAPUR
MANIKPUR
Paisuni River
Yamuna River
Bhadedu
Barduwara
INDEX
Contours showing postmonsoon rise and fall of water level (INM) during the year 1980-81.
Contours showing premonsoon rise and fall of water level (INM)-1980 81.
Contours showing premonsoon water level (INM) with respect to M.S.L. (1981).
Hydrograph station
Percentage of availability and utilisation of ground water.
C
RAMNAGAR
MANIKPUR
SHIVRAMPUR
WELL HYDROGRAPH
B
PAHARI
DEPTH OF WATER LEVEL (M)
AVERAGE RAINFALL IN CM
WELL N-1
SOURCE GROUND WATER INVESTIGATION ORGANISATION LUCKNOW
MONTHS (YEAR 1977)

Fig. 1.22

forming the rapids and waterfalls and the thick jungles provide the area with excellent scenic beauty which has great potentials for attracting tourists from all over the world. The great epic 'Ramayana' testifies clearly to the fact. Lord Rama chose Chitrakut to pass his time in exile, as a spot of grandeur, peace and abundant natural wealth. Many saints and sages also found the place praiseworthy and ideal for transcendental meditation etc. Goswami Tulsidas also could get a 'darshan' of 'Lord Rama' at this very spot. As a matter of fact, cultural eco-system clearly exhibits the religious impact of 'Ramayan' coupled with the natural landscape.

The northern plain regions near Yamuna and Paisuni are the recently developed agricultural lands where the development of irrigation facilities have offered some reliable cultivation. Single crop fields are still in majority. Water problem is very acute. Double cropping is possible only in the well irrigated areas. Soil is poor and so also the crops.

As society developed, man's impact on nature grew in scope and strength. However, in comparison to the state (U.P.), the Chitrakut region is most backward. The governments of Madhya Pradesh and Uttar Pradesh have drawn out plans for the development of 'Chitrakut Dham'. These schemes are bound to improve the socio-cultural landscape. However we should be very careful lest the nature is damaged and the restorative capabilities are progressively weakened and the human environment deteriorates affecting adversely the quality of life. We must preserve and improve the environment through ecological approach. The present settlement system consists of a few urban centres (Karwi, Sitapur, Rajapur, Pahari Buzurg and Bharatkup), the sparsely populated southern parts (patha area) and the relatively denser parts (average population density as 201 persons per sq. km.) in the plain region towards north. According to 1981 census the total population of the region is 518697 out of which 211311 or nearly 40.70 per cent area the workers. The topographical sheets register number of 590 villages in the entire study region wherein 93 villages are settled in southern upland region while the remaining villages are seated in plain and corresponding rolling surfaces. These records validate that the plateau region is thinly populated. The region has recorded about 28.67 per cent of population growth during the years 1971-1981. About 95 per cent of population is increased between the years 1951-1981 in the region. The population in the last two decades has increased due to general rise in economic conditions and less loss of life due to more medical facilities under the successive Five Year Plans. The social status of the farmers and agricultural workers are poorer. About 92.20 per cent of agricultural workers of the working population are registered in the region. The villages settled in plateau region feel acute problems of drinking water, food etc. while of the plain are relatively better in their standard of living. In plain region, about 36.12 per cent of the villages are electrified.

Apart from the above, medical facilities and educational infrastructure are very poor. About 1.14 primary health centres are recorded at per lakh population of the region (1980-1987). As a matter of literacy standard, the total literacy percentage exists as 18.04 in 1981 census. Male literacy (30.25%) is higher than that of female literacy (4.364%). On the basis of the statistical records of 1987, about 569 villages of the corresponding development blocks are provided with drinking water facilities by wells, handpipes and by other means. For irrigational purposes, there are lack of government tube-wells. Only two tube-wells are noticed in Naraini region but the number of private tube-wells is as high as 903.

The industrial activity is confined to some stone quarrying and crushing but has great potentialities in building stones and minerals like Bauxite, Agate, Lithomarge, and Silica sand. The dolomite boulders have also great economic value in view of cement industry. It is reported that the number of workers engaged in household industry is as 2653. Allahabad-Banda road and the Manikpur-Banda-Jhansi railway are the only lifeline providing transport facilities to the region. The roads are poor and less developed. Most of the areas are inaccessible. The average length (in km.) of P.W.D. roads at per lakh population is measured as 86.96 (1980-1986) while the average length (in km.) of P.W.D. metalled roads at per 1000 km^2 is 132.84. Thus the region is backward with respect to transport network.

The natural fauna and flora which once flourished in exhubrance are deteriorating fast in number as well as species. So we must adopt the ecological approach. We have to develop to ecological awareness among the masses and set a nice balance between man and environment.

RESEARCH DESIGN, STUDY ORIENTATION AND WORLDWIDE IMPORTANCE

Geomorphology is the study of science of landforms. Landforms are the product of the processes operating them. Thus, the geomorphology deals with the critical analysis of the geomorphological processes and its product; the assemblage of landforms. The subject relates principally with physical features of earth history. The processes and landforms and the so-called natural embalance have been deeply influenced by human interactions. Thus the culture cropping over the earth surface is cultivated by physical and human phenomena. The orientation of the study and researches was firstly directed towards the face of the earth showing physical attractions. The growing human activity has designated an unique glimpse of cultural landscapes. The creativity of man has mirrored before him, the all kind of necessities into possibilities. The researches and the burst of the publications came into the light on interactions between man and nature in the second phase of subject development.

In the age of science and technological revolution, man, from the face of the earth is making race towards space. He has destructed the nature for the new construction. Research advances in environmental monitoring and biological sciences combined with harsh experience in socio-economic development at the international level, have drawn the world community to an awareness that its own health and quality of life and the fate of future generations depend on action to avert environmental catastrophe. Thus the research design and orientation is turned over environmental management. This new understanding has helped to bring into sharper focus for the interrelatedness of environment.

Rapidly increasing population and comparatively the lack and unscientifically misuse of natural resources have caused serious problems for human civilization. The conservation of resources and the control of the environmental diversity are the major challenging questions. The eco-degradation among the developing countries is more acute. It is under control but dangerous between developed countries. The disturbed economic development through industrialisation at the cost of environment would ultimately lead to disaster. This will press the culture to eat, drink and breadth under pollution. These alarming causes enforced the man to do better for the civilization but he has little efforts for their construction comparatively achieving more for their destruction. Now the researches are going to settle basically for eco-protection not to earth description. The existing population and the future generation of this planet will certainly feel the necessity and importance of environment study.

PRECEDENCE AND PERFORMANCE OF FUTURE ECO-PROSPECTS

The Chitrakut region is a part and parcel of Vindhya mountains characterised by a vast stratified formation of sandstones, shales and limestone of Vindhyan system. It is less disturbed. The epeirogenic upheaval which lifted up the Vindhyan deposits from the floor of the sea to form a continental land area was the last serious earth movement recorded in the history of the peninsula. It remained an impassive solid block of the lithosphere without experiencing any folding or plication but proved the symptom of slight movement of secular upheaval and depression along their fringes and hence the sandstone country bespeaks the recital of shallow water deposition in their oft-recurring ripple marks and sun-cracked surfaces and in their conspicuous current bedding or diagonal lamination (Wadia, D.N. 1961). The different unconformities marked by conglomerates separating the different series are the exquisite notes of distinction. As a matter of present day scenery it is well defined by past morphological history. The plateau region is badly consumed by erosional processes and experiences senile stage of terrain characteristic. The most of the northern part has become peneplain while the land adjoining the plateau is in monadnock stage. The sloping ground is deeply ravaged by ravination processes where rills and ravines are most common. There is drinking as well as irrigational waters problem in the whole area. The measure of contour bunding and tree plantation are set in motion. It is believed that the region will feel more acute environmental problems in near future if it is not purposely surveyed in the sense of formulating the remedial processes.

PROPOSED RESEARCH PLAN

Apart from the above introductory approach, it is attempted to highlight the crux and clues of major morphological processes and resultant topography of Chitrakut Upland in the proposed research plan. The intensity and development procedure of weathering and fluvial processes and landforms are critically analysed through field survey and topographical sheets interpretation. A special analysis of major landforms like dolomite caves and tors have been taken into consideration. The existing environmental problems (deforestation, stone quarries, soil moisture and water deficit and the hazardous condition of ravines) its remedial processes and eco-development planning are also included in the last of the plan. The study be applied in sense of environmental geomorphology.

2

Background of Major Morphometric Properties

In order to provide basis for the study and interpretation of the morphological processes and landforms of the region, a general framework of major morphometric properties including linear, areal and relief properties of drainage network, major components of drainage hierarchy, slope dynamics, and drainage dissection have been introduced in this chapter. For this purpose, the twenty drainage basins covering the entire area and range have been selected and their basinal and regional characteristics have been critically examined. The analysis and the results are mirrored through quantitative techniques based on topographical sheets.

LINEAR PROPERTIES OF DRAINAGE NETWORK

Stream Ordering

Stream order is a major of a stream in the hierarchy of tributaries wherein the branching pattern of stream segments of all orders within a given watershed is analysed. Several geomorphologists like Horton, R.E. (1932, 1945), Strahler, A.N. (1964), Shreve, R.L. (1966, 1967), Scheidegger, A.E (1965, 1966), Smart, J.S. (1967, 1972) and Lewin, J. (1970) have proposed the techniques of stream ordering (Fig. 2.1). According to Strahler's stream segment method, the 20 drainage basins of the region have been segmented (Figs. 2.2 and 2.3) and the number of tributaries (segments) according to orders have been listed in Table 2.1. The total number of stream segments of different orders in 20 drainage basins are as 1583 (Table 2.1) wherein 1210, 281, 69, 19 and 04 streams have been noted in I, II, III, IV and V orders respectively. Samrani, Paroi, Piprawal and Kachchua basins generally emerge over the alluvium flat country while Barwa, Kewai, Jhurai, Gauhia, Dhora and Dingchi basins predominate over plateau (upper reaches) and flat plain terrain (lower reaches). The remaining basins (Ghinhal, Jhuri, Bagdara, Kalibarah, Maro, Sukra, Kulagara, Pindrah, Barar and Ganta) generally drain over plateau and hilly areas.

Table 2.1: Numbers of stream segments (N_u)

Sl. No.	*Drainage basins*	*Orders (U)*					*Total Nos. of stream segments (N_u)*
		1	*2*	*3*	*4*	*5*	
		N_{u_1}	N_{u_2}	N_{u_3}	N_{u_4}	N_{u_5}	
1.	Samrani	34	06	01	–	–	41
2.	Paroi	39	07	02	01	–	49
3.	Piprawal	32	07	02	01	–	42
4.	Barwa	168	45	09	02	01	225
5.	Ghinhal	119	27	07	01	–	154
6.	Jhuri	76	14	04	01	–	95
7.	Bagdara	70	17	04	01	–	92
8.	Kalibarah	105	24	07	02	01	139
9.	Maro	72	17	03	01	–	93
10.	Sukra	46	14	03	01	–	64
11.	Kulagara	41	09	02	01	–	53
12.	Pindrah	42	10	01	–	–	53
13.	Barar	63	18	05	02	01	89
14.	Kewai	145	34	08	03	01	191
15.	Ganta	49	11	03	01	–	64
16.	Jhurai	19	04	01	–	–	24
17.	Gauhia	18	03	01	–	–	22
18.	Kachchua	18	02	01	–	–	21
19.	Dhora	22	05	01	–	–	28
20.	Dingchi	32	11	04	01	–	48
Total number of stream segments		1210	281	69	19	04	1583

Bifurcation Ratio (R_b)

Horton, R.E. (1932) defined 'Bifurcation Ratio' (R_b) as the relationship between the number of stream segments of a given order (N_u) to the number of stream segments of the next order which may be represented as:

$$R_b = \frac{N_u}{N_u + 1}$$

where, R_b = Bifurcation Ratio

N_u = number of stream segments of a given order

and $N_u + 1$ = number of stream segments of the next higher orders.

Bifurcation ratio is a dimensionless property of an hierarchical system of a drainage basin. It is basically controlled by the physical eco-system.

The bifurcation ratio of 20 drainage basins has been calculated with the help of the Horton's formula. Mean bifurcation ratio of all the drainage basins (Table 2.2) ranges between 3.25 (Dingchi basin) and 7.10 (Pindrah basin). Table 2.2 and Fig. 2.4A illustrates that the values of bifurcation ratios generally decrease with increasing orders of the basins. Basins like Samrani, Dhora, Gauhia and Jhurai show increasing trends of bifurcation ratios with increasing areal coverages (Fig. 2.5). Except a few basins of 4th and 5th orders, mostly the drainage basins depict the increasing trends of R_b with their increasing areas (Table 2.2). It is found stable on uniform lithology and vary region to region (Fig. 2.6) with respect to structural control (Prasad, G. 1986).

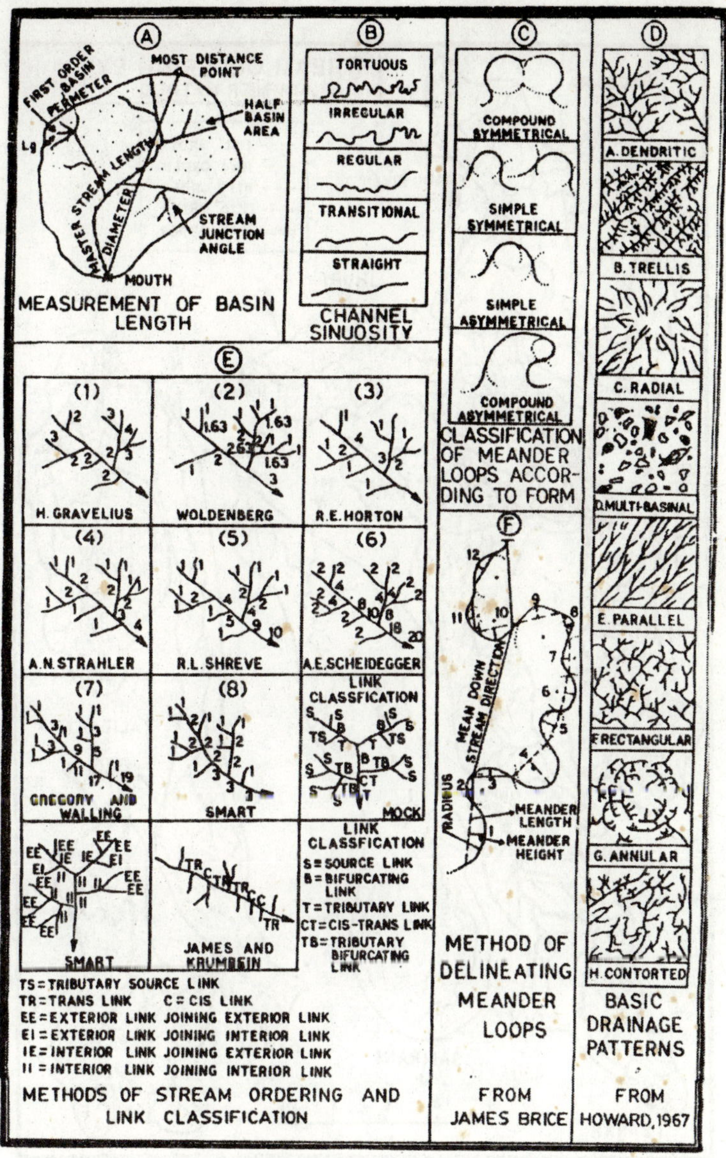
A
MOST DISTANCE POINT
FIRST ORDER BASIN PERIMETER
HALF BASIN AREA
Lg
MASTER STREAM LENGTH
DIAMETER
STREAM JUNCTION ANGLE
MOUTH
MEASUREMENT OF BASIN LENGTH
B
TORTUOUS
IRREGULAR
REGULAR
TRANSITIONAL
STRAIGHT
CHANNEL SINUOSITY
C
COMPOUND SYMMETRICAL
SIMPLE SYMMETRICAL
SIMPLE ASYMMETRICAL
COMPOUND ASYMMETRICAL
CLASSIFICATION OF MEANDER LOOPS ACCORDING TO FORM
D
A. DENDRITIC
B. TRELLIS
C. RADIAL
D. MULTI-BASINAL
E. PARALLEL
F. RECTANGULAR
G. ANNULAR
H. CONTORTED
BASIC DRAINAGE PATTERNS
FROM HOWARD, 1967
E
(1) H. GRAVELIUS
(2) WOLDENBERG
(3) R.E. HORTON
(4) A.N. STRAHLER
(5) R.L. SHREVE
(6) A.E. SCHEIDEGGER
(7) GREGORY AND WALLING
(8) SMART
LINK CLASSFICATION
MOCK
SMART
JAMES AND KRUMBEIN
LINK CLASSFICATION
S = SOURCE LINK
B = BIFURCATING LINK
T = TRIBUTARY LINK
CT = CIS-TRANS LINK
TB = TRIBUTARY BIFURCATING LINK
TS = TRIBUTARY SOURCE LINK
TR = TRANS LINK C = CIS LINK
EE = EXTERIOR LINK JOINING EXTERIOR LINK
EI = EXTERIOR LINK JOINING INTERIOR LINK
IE = INTERIOR LINK JOINING EXTERIOR LINK
II = INTERIOR LINK JOINING INTERIOR LINK
METHODS OF STREAM ORDERING AND LINK CLASSIFICATION
F
MEAN DOWN STREAM DIRECTION
RADIUS
MEANDER LENGTH
MEANDER HEIGHT
METHOD OF DELINEATING MEANDER LOOPS
FROM JAMES BRICE

Fig. 2.1

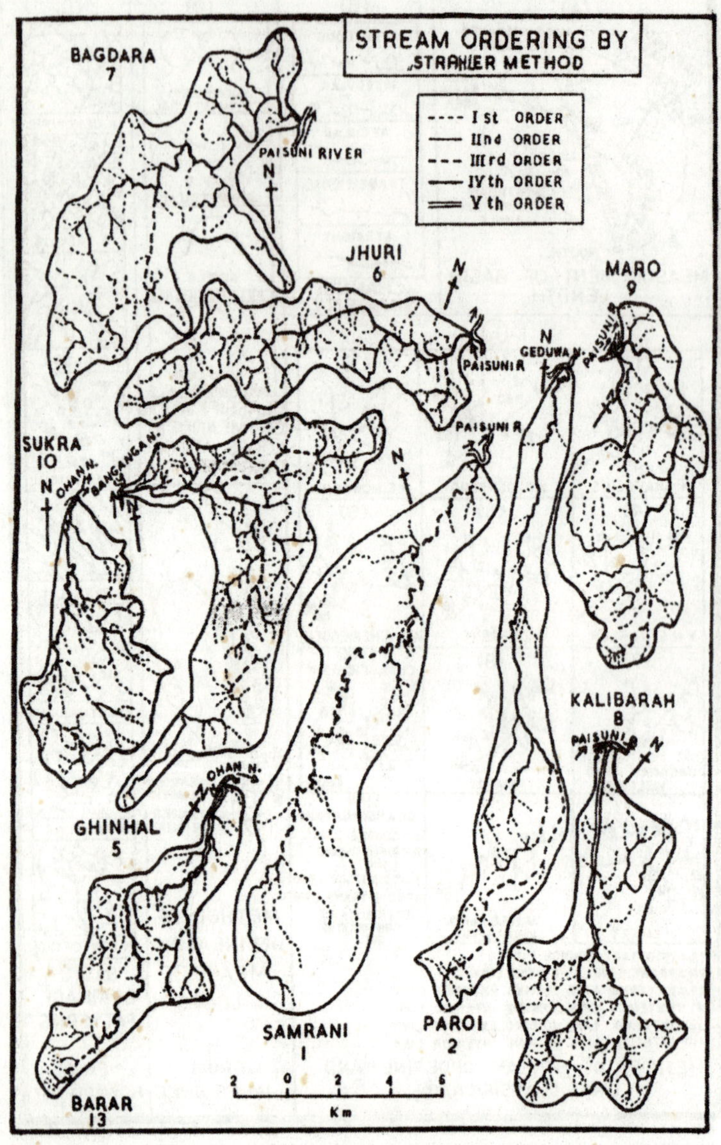
STREAM ORDERING BY
STRAHLER METHOD
I st ORDER
IInd ORDER
IIIrd ORDER
IVth ORDER
Vth ORDER
BAGDARA
7
PAISUNI RIVER
N
JHURI
6
N
PAISUNI R
MARO
9
N
GEDUWA N.
PAISUNI R
PAISUNI R
SUKRA
10
N
OHAN N.
BANGANGA N.
N
KALIBARAH
8
PAISUNI R
N
OHAN N.
GHINHAL
5
SAMRANI
1
PAROI
2
2 0 2 4 6
Km
BARAR
13

Fig. 2.2

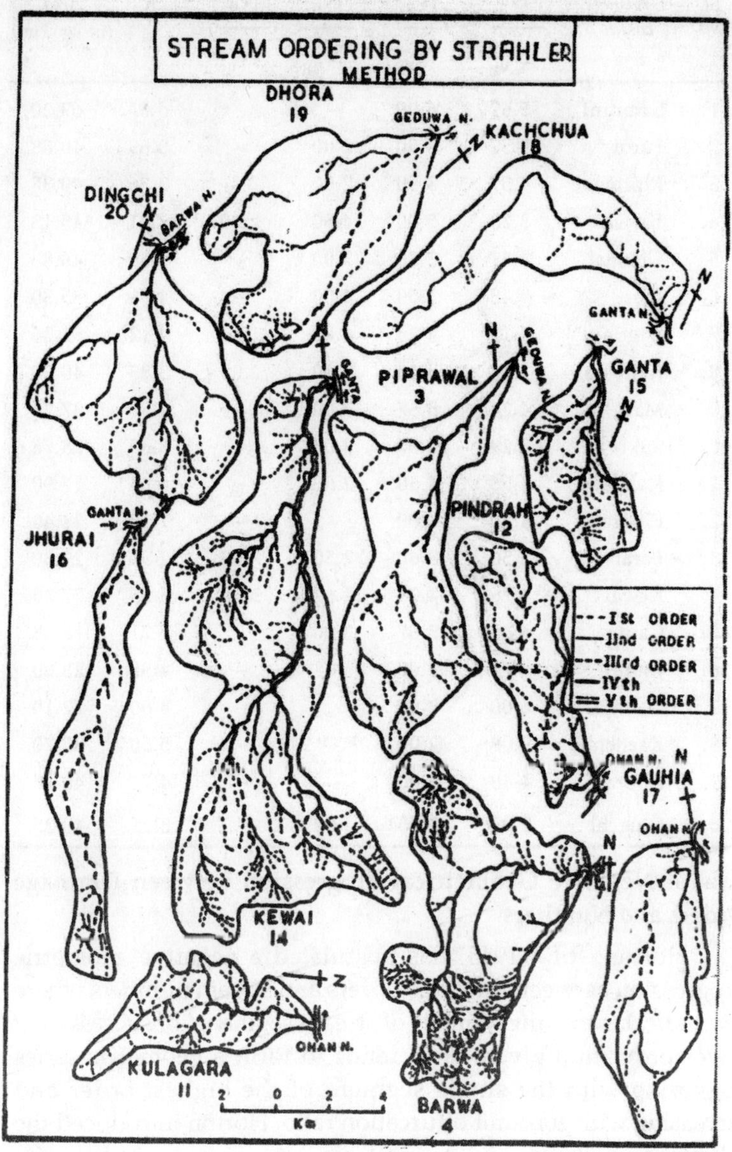
STREAM ORDERING BY STRAHLER METHOD
DHORA 19
GEDUWA N.
KACHCHUA 18
DINGCHI 20
BARWA N.
GANTA N.
GANTA
PIPRAWAL 3
GEDUWA N.
GANTA 15
PINDRAH 12
GANTA N.
JHURAI 16
I St ORDER
IInd ORDER
IIIrd ORDER
IVth
Vth ORDER
OHAN N.
GAUHIA 17
OHAN N.
KEWAI 14
OHAN N.
KULAGARA 11
2 0 2 4
Km
BARWA 4

Fig. 2.3

Table 2.2: Bifurcation ratio (R_b)

Sl. No.	*Drainage basins*	$\frac{N_{u_1}}{N_{u_2}}$	$\frac{N_{u_2}}{N_{u_3}}$	$\frac{N_{u_3}}{N_{u_4}}$	$\frac{N_{u_4}}{N_{u_5}}$	R_b	*Area (in sq. km.)*
1.	Samrani	5.67	6.00	–	–	5.84	67.20
2.	Paroi	5.57	3.50	2.00	–	3.69	40.83
3.	Piprawal	4.57	3.50	2.00	–	3.36	40.38
4.	Barwa	3.73	5.00	4.50	2.00	3.81	45.13
5.	Ghinhal	4.41	3.86	7.00	–	5.09	46.83
6.	Jhuri	5.43	3.50	4.00	–	4.31	35.30
7.	Bagdara	4.12	4.25	4.00	–	4.12	52.25
8.	Kalibarah	4.38	3.43	3.50	2.00	3.33	40.38
9.	Maro	4.24	5.67	3.00	–	4.30	37.73
10.	Sukra	3.29	4.67	3.00	–	3.65	28.78
11.	Kulagara	4.56	4.50	2.00	–	3.69	17.90
12.	Pindrah	4.20	10.00	–	–	7.10	21.40
13.	Barar	3.50	3.60	2.50	2.00	2.90	29.30
14.	Kewai	4.26	4.25	2.67	3.00	3.55	77.35
15.	Ganta	4.45	3.67	3.00	–	3.71	16.48
16.	Jhurai	4.75	4.00	–	–	4.38	23.00
17.	Gauhia	6.00	3.00	–	–	4.50	32.10
18.	Kachchua	9.00	2.00	–	–	5.50	35.70
19.	Dhora	4.40	5.00	–	–	4.70	40.15
20.	Dingchi	2.99	2.75	4.00	–	3.25	40.25

Law of Negative Geometrical Progression between Drainage Orders and Numbers

Horton, R.E. (1945) has postulated a negative geometric progression between drainage orders and stream numbers where law indicates that the number of stream segments of successively lower order in a given basin tends to form a geometric series beginning with the single segment of the highest order and increasing with constant bifurcation ratio. Horton introduced the following equation:

$N_u = R_b (K-u)$

where

N_u = number of stream segments of a given order;

R_b = constant bifurcation ratio; and

K = highest order of the basin.

The above conditions are not possible in nature because none of the basin may exhibit constant R_b. Thus, the regression coefficients have been shown (Fig. 2.7) for each basin to show the relationship between orders and stream segments adopting the formula as suggested by Strahler (1964):

$$\text{Log } y = \text{Log } a - bx$$

where

y = number of stream segments (N_u),

x = stream order (u),

a = constant, and

b = regression coefficient.

On the basis of the above equation, regression lines of negative exponential functions have been plotted against stream orders and number of stream segments for 20 drainage basins of the region (Fig. 2.7). It is remarkable that the above model holds good in the case of 3rd and 4th orders of the basins. Few examples of departure are also marked which are cleared with coefficient of correlation. The above discussion concludes that the lower order receives the larger number of stream segments in all the basins. It is also apparent that the youngest basins of the lower order usually confirm closely to the regression lines than that of the basins of higher order and older in age (Prasad, G. 1987).

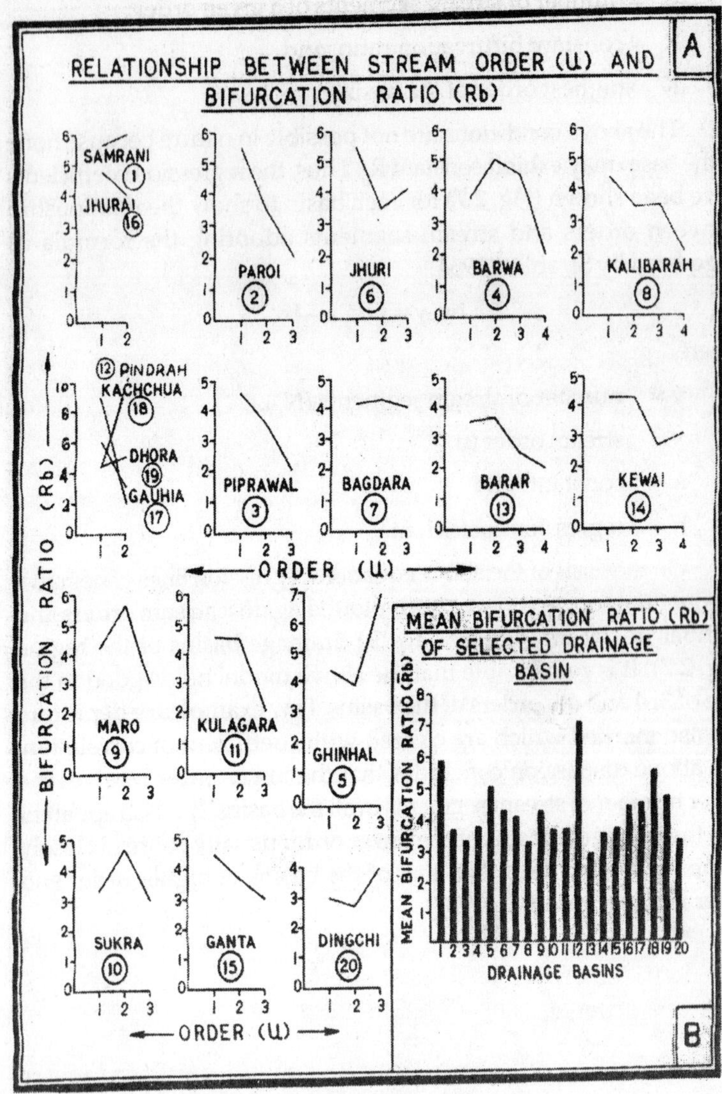
A
RELATIONSHIP BETWEEN STREAM ORDER (U) AND BIFURCATION RATIO (Rb)
SAMRANI 1
JHURAI 16
PAROI 2
JHURI 6
BARWA 4
KALIBARAH 8
12 PINDRAH
KACHCHUA 18
DHORA 19
GAUHIA 17
PIPRAWAL 3
BAGDARA 7
BARAR 13
KEWAI 14
ORDER (U)
BIFURCATION RATIO (Rb)
MARO 9
KULAGARA 11
GHINHAL 5
SUKRA 10
GANTA 15
DINGCHI 20
ORDER (U)
MEAN BIFURCATION RATIO (Rb) OF SELECTED DRAINAGE BASIN
MEAN BIFURCATION RATIO (Rb)
DRAINAGE BASINS
B

Fig. 2.4

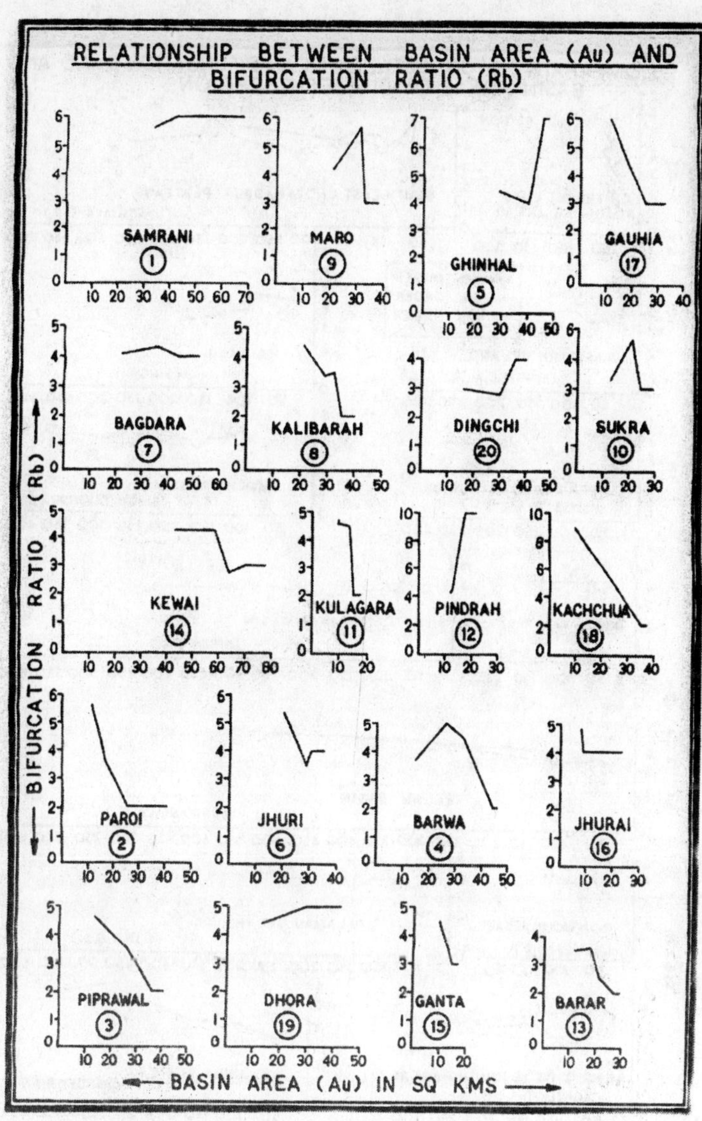
RELATIONSHIP BETWEEN BASIN AREA (Au) AND BIFURCATION RATIO (Rb)
SAMRANI
1
MARO
9
GHINHAL
5
GAUHIA
17
BAGDARA
7
KALIBARAH
8
DINGCHI
20
SUKRA
10
KEWAI
14
KULAGARA
11
PINDRAH
12
KACHCHUA
18
PAROI
2
JHURI
6
BARWA
4
JHURAI
16
PIPRAWAL
3
DHORA
19
GANTA
15
BARAR
13
BIFURCATION RATIO (Rb)
BASIN AREA (Au) IN SQ KMS

Fig. 2.5

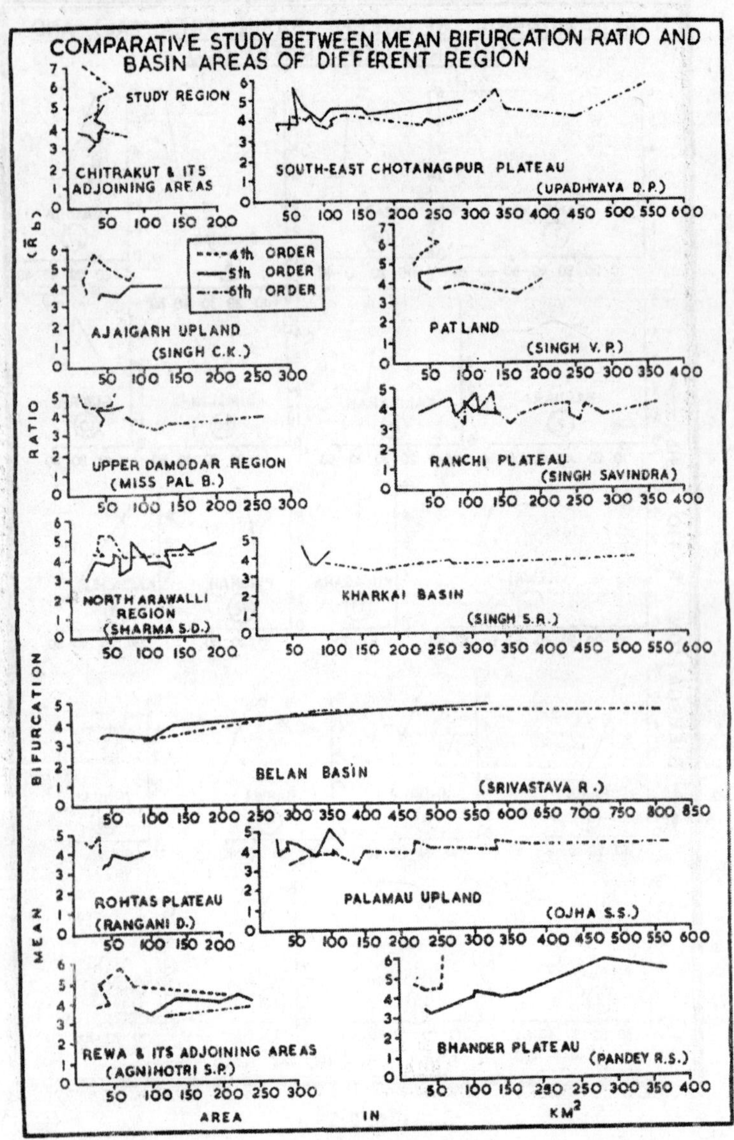

COMPARATIVE STUDY BETWEEN MEAN BIFURCATION RATIO AND BASIN AREAS OF DIFFERENT REGION
STUDY REGION
CHITRAKUT & ITS ADJOINING AREAS
SOUTH-EAST CHOTANAGPUR PLATEAU
(UPADHYAYA D.P.)
4th ORDER
5th ORDER
6th ORDER
AJAIGARH UPLAND
(SINGH C.K.)
PAT LAND
(SINGH V. P.)
UPPER DAMODAR REGION
(MISS PAL B.)
RANCHI PLATEAU
(SINGH SAVINDRA)
NORTH ARAWALLI REGION
(SHARMA S.D.)
KHARKAI BASIN
(SINGH S.R.)
BELAN BASIN
(SRIVASTAVA R.)
ROHTAS PLATEAU
(RANGANI D.)
PALAMAU UPLAND
(OJHA S.S.)
REWA & ITS ADJOINING AREAS
(AGNIHOTRI S.P.)
BHANDER PLATEAU
(PANDEY R.S.)
MEAN BIFURCATION RATIO (R̄ b)
AREA IN KM²

Fig. 2.6

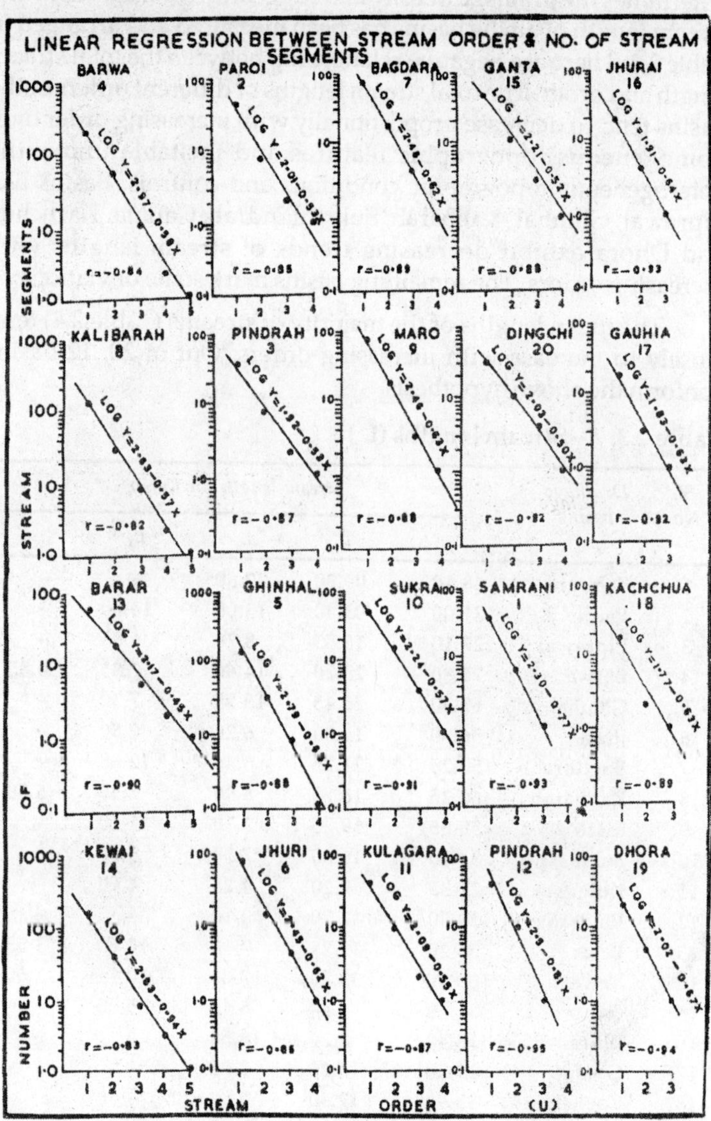
LINEAR REGRESSION BETWEEN STREAM ORDER & NO. OF STREAM SEGMENTS
BARWA 4
PAROI 2
BAGDARA 7
GANTA 15
JHURAI 16
KALIBARAH 8
PINDRAH 3
MARO 9
DINGCHI 20
GAUHIA 17
BARAR 13
GHINHAL 5
SUKRA 10
SAMRANI 1
KACHCHUA 18
KEWAI 14
JHURI 6
KULAGARA 11
PINDRAH 12
DHORA 19
NUMBER OF STREAM SEGMENTS
STREAM ORDER (U)

Fig. 2.7

Length Ratio

Stream length is a significant morphometric parameter which determines the drainage density and ruggedness of the terrain. The stream length of all the basins has been measured and arranged in Table 2.3. There are negative relationship between the total stream length and order. The total stream lengths of different orders of the basins tend to decrease proportionally with increasing order over homogeneous topographic features and unstable under the heterogeneous topographic conditions and controls. Basins like Piprawal, Ghinhal, Kalibarah, Sukra, Pindrah, Gauhia, Kachchua and Dhora exhibit decreasing trends of stream lengths with increasing orders. The remaining basins mark some deviations.

The mean lengths of the majority of streams (Table 2.4) tend closely to increase with increasing orders. Out of 20, 12 basins conform the above hypothesis.

Table 2.3: Stream lengths (L_u)

Sl. No.	*Drainage basins*	*Stream length (in Kms.)*				
		L_1	L_2	L_3	L_4	L_5
1.	Samrani	48.10	09.30	22.80	–	–
2.	Paroi	31.00	05.40	13.00	14.80	–
3.	Piprawal	27.10	15.55	8.70	6.85	–
4.	Barwa	75.80	25.20	14.60	17.35	0.85
5.	Ghinhal	67.90	25.45	15.75	7.30	–
6.	Jhuri	56.30	14.00	6.20	9.80	–
7.	Bagdara	50.25	21.25	8.40	12.65	–
8.	Kalibarah	60.25	18.70	8.00	7.55	6.50
9.	Maro	50.85	19.25	3.10	11.10	–
10.	Sukra	34.40	16.60	7.50	5.70	–
11.	Kulagara	27.35	7.20	4.25	4.60	–
12.	Pindrah	30.50	12.70	10.10	–	–
13.	Barar	36.20	12.25	7.10	10.60	3.55
14.	Kewai	103.10	33.70	17.00	10.75	11.00
15.	Ganta	31.20	6.40	5.45	5.60	–
16.	Jhurai	18.40	2.20	18.50	–	–
17.	Gauhia	19.50	15.20	7.25	–	–
18.	Kachchua	13.90	17.30	0.50	–	–
19.	Dhora	22.40	16.50	10.70	–	–
20.	Dingchi	30.20	17.50	5.50	5.70	–

Table 2.4: Mean lengths (L_u) of stream segments (in Kms.)

Sl. No.	Drainage basins	Drainage orders' (U)				
		L_1	L_2	L_3	L_4	L_5
1.	Samrani	1.41	1.55	22.80	–	–
2.	Paroi	0.795	0.771	6.50	14.80	–
3.	Piprawal	0.85	2.22	4.35	6.85	–
4.	Parwa	0.45	0.56	1.62	8.68	0.85
5.	Ghinhal	0.570	0.94	2.25	7.30	–
6.	Jhuri	0.74	1.00	1.55	9.80	–
7.	Bagdara	0.72	1.25	2.10	12.65	–
8.	Kalibarah	0.573	0.78	1.14	3.78	6.50
9.	Maro	0.71	1.13	1.03	11.10	–
10.	Sukra	0.75	1.19	2.50	5.70	–
11.	Kulagara	0.67	0.80	2.13	4.60	–
12.	Pindrah	0.73	1.27	10.10	–	–
13.	Barar	0.574	0.68	1.42	5.30	3.55
14.	Kewai	0.711	0.991	2.13	3.58	3.55
15.	Ganta	0.64	0.58	1.82	5.60	–
16.	Jhurai	0.97	0.55	18.50	–	–
17.	Gauhia	1.08	5.07	7.25	–	–
18.	Kachchua	0.772	8.65	0.50	–	–
19.	Dhora	1.02	3.30	10.70	–	–
20.	Dingchi	0.94	1.59	1.38	5.70	–

Length ratio (R_L) is termed as the ratio between the mean lengths of stream segments of two successive orders and is derived according to the following formula (Horton, R.E. 1945):

$$R_L = \frac{\bar{L}_u}{L_u^{-1}}$$

when

$$\bar{L}_u = \frac{L_u}{N_u}$$

where

$\bar{L}_u$ = mean stream length of all segments of a given order,

L_u = total stream length of all segments of a given order, and

N_u = number of stream segments of a given order.

The cumulative mean lengths of all the basins have been given in Table 2.5. The mean length ratio of 20 sample basins is also shown in Table 2.6. It varies between 1.92 and 17.10. Significant variations of length ratio between one basin to others and even in a single basin (between one order and the next) are well marked. This wide range of variations are due to the topographical change and lithological controls. Horton estimates that the values of the length ratios should range between 2 and 3. In the present case only ten basins confirm the Hortonian law.

It is apparent from Table 2.6 that the highest order basins and older in age like Barwa, Kalibarah, Barar and Kewai record lowest length ratio whereas the lowest order basins (Samrani, Pindrah, Jhurai, Gauhia, Kachchua and Dhora) and comparatively younger in age exhibit highest mean length ratio. Jhurai (R_L = 17.10 km.) a third order basin shows peculiar result. The highest length ratio is also noted in Paroi, Jhuri, Bagdara and Maro drainage basins.

Table 2.5: Cumulative mean lengths (in kms.)

Sl. No.	*Drainage basins*	*Drainage orders (U)*				
		1	2	3	4	5
1.	Samrani	1.41	2.96	25.76	–	–
2.	Paroi	0.795	1.57	8.07	22.87	–
3.	Piprawal	0.85	3.07	7.42	14.27	–
4.	Barwa	0.45	1.01	2.63	11.31	12.16
5.	Ghinhal	0.57	1.51	3.76	11.06	–
6.	Jhuri	0.74	1.74	3.29	13.09	–
7.	Bagdara	0.72	1.97	4.07	16.72	–
8.	Kalibarah	0.57	1.35	2.49	6.27	12.77
9.	Maro	0.71	1.84	2.87	13.97	–
10.	Sukra	0.75	1.93	4.43	10.13	–
11.	Kulagara	0.67	1.47	3.59	8.19	–
12.	Pindrah	0.73	2.00	12.10	–	–
13.	Barar	0.57	1.25	2.67	7.97	11.52
14.	Kewai	0.71	1.70	3.83	7.41	18.41
15.	Ganta	0.63	1.22	3.03	8.63	–
16.	Jhurai	0.97	1.52	20.02	–	–
17.	Gauhia	1.08	6.15	13.40	–	–
18.	Kachchua	0.77	9.42	9.92	–	–
19.	Dhora	1.02	4.32	15.02	–	–
20.	Dingchi	0.94	2.53	3.91	9.61	–

Table 2.6: Length ratio (R_L)

Sl. No.	*Drainage basins*	L_2/L_1	L_3/L_2	L_4/L_3	L_5/L_4	$\bar{R}_L$
1.	Samrani	1.10	14.70	–	–	7.90
2.	Paroi	0.97	8.43	2.28	–	3.89
3.	Piprawal	2.62	1.96	1.57	–	2.05
4.	Barawa	1.24	2.89	5.35	0.098	2.39
5.	Ghinhal	1.65	2.39	3.24	–	2.43
6.	Jhuri	1.35	1.55	6.32	–	3.07
7.	Bagdara	1.74	1.68	6.02	–	3.15
8.	Kalibarah	1.36	1.46	3.31	1.71	1.96
9.	Maro	1.60	0.91	10.78	–	4.43
10.	Sukra	1.59	2.11	2.28	–	1.99
11.	Kulagara	1.20	2.66	2.16	–	2.01
12.	Pindrah	1.75	7.95	–	–	4.85
13.	Barar	1.18	2.09	3.73	0.67	1.92
14.	Kewai	1.39	2.14	1.69	3.07	2.07
15.	Ganta	0.91	3.13	3.08	–	2.37
16.	Jhurai	0.57	33.64	–	–	17.10
17.	Gauhia	4.69	1.43	–	–	3.06
18.	Kachchua	11.20	0.06	–	–	5.63
19.	Dhora	3.24	3.24	–	–	3.24
20.	Dingchi	1.69	0.87	4.13	–	2.23

On the basis of the above discussion, it may be inferred that younger the river, the more abnormal development of a stream length and also the higher abnormality in length ratio are observed. On the contrary older the river, the more normal development of stream lengths and thus lesser abnormality in length ratios exist.

It is obvious that the length ratio between different orders of the basin do not follow any rule of either consistent decreases or increase with orders. As such, it becomes independent of drainage orders.

Law of Positive Geometric Progression of Stream Length

Horton, R.E. (1945) second law of drainage composition is related to the law of stream lengths which states that the cumulative mean lengths of stream segments of successive higher orders increase in geometrical progression starting with the mean length of the 1st order segments with constant length ratio. It is indicated by the following positive exponential function model:

$$L_u = L_1 R_L (u-1)$$

where

$u =$ the order of the stream;

$L_1 =$ the mean length of the 1st order streams; and

$R_L =$ is the length ratio.

The above law has been examined through the straight regression lines of positive function model based on the following regression equation:

$$\text{Log } y = \text{Log } a + bx$$

where

$y =$ cumulative mean length;

$x =$ order of the streams;

$a =$ constant; and

$b =$ regression coefficient.

The regression lines (Fig. 2.8) are drawn on the basis of the above equation to test the law of stream length of positive exponential function. It reveals the fact that there is linear relationship between orders and cumulative mean lengths as observed in the case of Pindrah and Dhora (III order), Bagdara, Ganta, Ghinhal, Sukra and Kulagara (IV order) and Kalibarah and Kewai (V order) basins where nearly all the points have come closer to the best fit lines. The remaining basins show some deviations due to topographic controls. The coefficient of correlation range between + 0.73 (Samrani basin) and + 0.99 (Gauhia basin) which also indicate strong positive correlation.

Generally the highest 'r' values of the highest order basins and the lowest 'r' value of the lowest order basins also indicate that older river develops their stream segments more normally than the younger river and thus the above law confirms more closely with older river than that of the younger rivers.

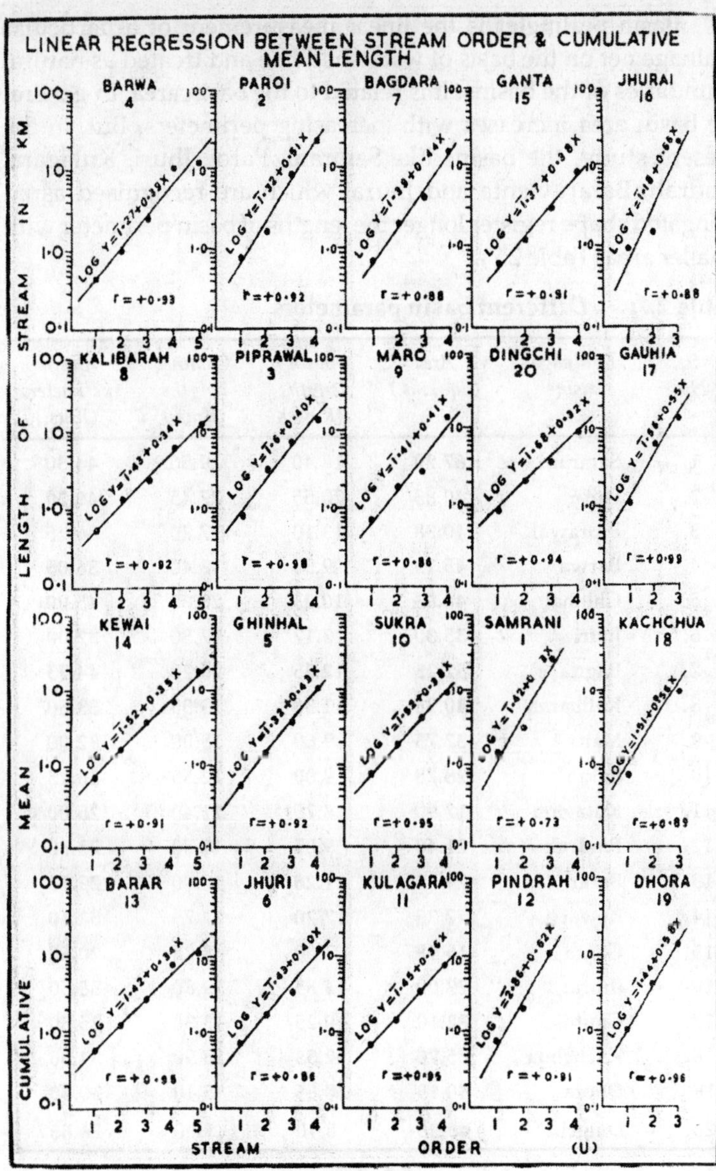
LINEAR REGRESSION BETWEEN STREAM ORDER & CUMULATIVE MEAN LENGTH
BARWA 4
LOG Y=1̄·27+0·39X
r=+0·93
PAROI 2
LOG Y=1̄·32+0·51X
r=+0·92
BAGDARA 7
LOG Y=1̄·38+0·44X
r=+0·88
GANTA 15
LOG Y=1̄·37+0·38X
r=+0·91
JHURAI 16
LOG Y=1̄·18+0·66X
r=+0·88
KALIBARAH 8
LOG Y=1̄·43+0·34X
r=+0·92
PIPRAWAL 3
LOG Y=1̄·6+0·40X
r=+0·98
MARO 9
LOG Y=1̄·41+0·41X
r=+0·86
DINGCHI 20
LOG Y=1̄·68+0·32X
r=+0·94
GAUHIA 17
LOG Y=1̄·56+0·55X
r=+0·99
KEWAI 14
LOG Y=1̄·52+0·35X
r=+0·91
GHINHAL 5
LOG Y=1̄·32+0·43X
r=+0·92
SUKRA 10
LOG Y=1̄·52+0·38X
r=+0·95
SAMRANI 1
r=+0·73
KACHCHUA 18
r=+0·89
BARAR 13
LOG Y=1̄·42+0·34X
r=+0·98
JHURI 6
LOG Y=1̄·43+0·40X
KULAGARA 11
LOG Y=1̄·45+0·36X
r=+0·95
PINDRAH 12
LOG Y=2̄·86+0·62X
r=+0·91
DHORA 19
LOG Y=1̄·44+0·58X
r=+0·96
CUMULATIVE MEAN LENGTH OF STREAM IN KM
STREAM ORDER (U)

Fig. 2.8

AREAL PROPERTIES OF DRAINAGE NETWORK

Basin Perimeters

Basin perimeter is the linear measurement of a particular drainage net on the basis of watershed line and treated as natural boundaries of the basins. It is related to the basin area. In general the basin area increases with increasing perimeters. But, in the present study, the basins like Samrani, Paroi, Jhuri, Kulagara, Pindrah, Barar, Ganta and Jhurai which are recognised as an elongated shape register longer the lengths of basin perimeter with smaller area (Table 2.7).

Table 2.7: Different basin parameters

S. No.	*Drainage basins*	*Area (Sq. kms.)*	*Basin lengths (Kms.)*	*Channel lengths (Kms.)*	*Basin perimeters (Kms.)*
1.	Samrani	67.20	18.40	29.50	44.50
2.	Paroi	40.83	20.55	27.45	46.20
3.	Piprawal	40.38	13.10	17.20	36.25
4.	Barwa	45.13	9.50	13.40	36.05
5.	Ghinhal	46.83	10.25	20.00	45.00
6.	Jhuri	35.30	13.17	17.50	37.00
7.	Bagdara	52.25	12.25	20.20	44.75
8.	Kalibarah	40.38	11.30	13.55	33.60
9.	Maro	37.73	9.60	15.00	32.00
10.	Sukra	28.78	9.00	12.55	26.25
11.	Kulagara	17.90	8.75	11.40	25.00
12.	Pindrah	21.40	9.05	11.80	23.80
13.	Barar	29.30	11.25	17.90	29.55
14.	Kewai	77.35	17.20	22.75	52.50
15.	Ganta	16.48	7.00	8.25	20.60
16.	Jhurai	23.00	14.85	19.80	38.50
17.	Gauhia	32.10	10.25	13.30	25.50
18.	Kachchua	35.70	12.55	18.90	34.00
19.	Dhora	40.15	10.55	15.10	30.50
20.	Dingchi	40.25	8.10	9.00	26.85

Basin Shape

The ideal drainage basin is usually of pear shape but since it is dependent on the size and length of the master stream of the basin and basin perimeter which are themselves dependent on other variables. On an average, three sub-categories of basin shape have been recognised viz. circular, elongated and idented. Although various methods for the measurement of shape indices have been suggested but the following four methods are applied in the present study:

1. Horton's (1932) Form Factor (F)

$$F = \frac{A}{L^2}$$

where A = basin, area, and

L = basin length.

2. Miller's (1953) Circularity Index (C)

$$C = \frac{4\pi A}{P^2}$$

where P = basin perimeter

3. Schumm's (1956) Elongation Ratio (E)

$$E = \frac{\text{Diameter of circle with same area as basin}}{\text{basin length}}$$

$$= \sqrt[2]{\frac{A / \Pi}{L}}$$

4. Chorley, Malm and Pogorzeiski (1957) Lemniscate Method (K)

$$K = \frac{L^2}{4A}$$

It is notable that the higher values of circularity index indicate the more circular shape of the basin while the lower the values of 'R' and 'F' and higher the values of 'K' indicate the more elongated shape of the basins.

On the basis of the above formula the derived results are arranged in Table 2.8. The critical analysis reveals the fact that the two shape indices viz. form factor and lemniscate 'K' yield more satisfactory results and more or less closer to reality (Prasad, G. 1987).

Table 2.8: Basin-shape indices

S. No.	*Drainage basins*	*Circularity index (CI)**	*Elongation ratio (R)++*	*Form factor (F)Q*	*Liminiscate (K)≠≠*
1.	Samrani	0.43	0.50	0.20	1.26
2.	Paroi	0.24	0.35	0.97	2.59
3.	Piprawal	0.39	0.43	0.24	1.06
4.	Barwa	0.44	0.80	0.50	0.50
5.	Ghinhal	0.29	0.75	0.45	0.56
6.	Jhuri	0.32	0.49	0.19	1.34
7.	Bagdara	0.33	0.66	0.35	0.72
8.	Kalibarah	0.45	0.63	0.32	0.79
9.	Maro	0.46	0.72	0.41	0.61
10.	Sukra	0.53	0.67	0.36	0.70
11.	Kulagara	0.36	0.54	0.23	1.07
12.	Pindrah	0.47	0.58	0.26	0.96
13.	Barar	0.42	0.54	0.23	1.08
14.	Kewai	0.35	0.58	0.26	0.96
15.	Ganta	0.49	0.65	0.34	0.74
16.	Jhurai	0.20	0.36	0.10	2.40
17.	Guahia	0.62	0.62	0.31	0.82
18.	Kachchua	0.39	0.54	0.23	1.10
19.	Dhora	0.54	0.68	0.36	0.69
20.	Kingchi	0.70	0.88	0.61	0.41

* The higher the value, the more circular shape;

++ The lower the value, the more elongated shape;

Q The lower the value, the more elongated shape; and

≠≠ The higher the value, the more elongated shape.

Law of Basin Area

It is appropriate to point out that the basin area becomes cumulative from 1st order to the successively higher orders and thus the area of a higher order (the entire basin area) stream includes the total area covered by all the streams of lower order. There are positive relationship between stream order and basin area. The basin area increases with an increases in order of stream segments (Table 2.9).

Table 2.9: Basin areas (Au) (in Sq. Kms.)

S. No.	*Drainage basins*	Au_1	Au_2	Au_3	Au_4	Au_5
1.	Samrani	33.38	42.33	67.20	–	–
2.	Paroi	11.03	17.18	24.93	40.83	–
3.	Piprawal	12.85	25.98	35.55	40.38	–
4.	Barwa	16.48	26.13	32.68	44.70	45.13
5.	Ghinhal	28.58	39.63	45.70	46.83	–
6.	Jhuri	20.48	27.58	30.05	35.30	–
7.	Bagdara	22.28	35.73	44.18	52.25	–
8.	Kalibarah	20.85	28.88	32.48	35.60	40.38
9.	Maro	21.05	31.13	32.65	37.73	–
10.	Sukra	12.80	20.90	23.83	28.78	–
11.	Kulagara	9.60	13.48	15.45	17.90	–
12.	Pindrah	11.90	17.28	21.40	–	–
13.	Barar	11.65	17.48	21.30	27.07	29.30
14.	Kewai	42.65	58.08	64.03	69.58	77.35
15.	Ganta	10.30	12.60	14.60	16.48	–
16.	Jhurai	7.75	8.35	23.00	–	–
17.	Gauhia	11.35	24.60	32.10	–	–
18.	Kachchua	10.08	35.38	35.70	–	–
19.	Dhora	13.00	29.00	40.15	–	–
20.	Dingchi	16.25	30.13	35.58	40.25	–

On the basis of Table 2.9 it may be pertinent to point out that the basins which are developed over highly dissected hilly terrain having much relief irregularities are higher in their order but smaller in their areal coverages comparatively those basins which drain over flat surfaces or partly over highly dissected hilly terrain and partly over flat country. Thus, relief regularity accelerates the areal coverages but minimizes the stream segments with their orders.

Area Ratio

The area ratio (R_a) is the proportion of increase of mean basin area (A_u) between two successive orders of a given basin. It is derived on the basis of the following formula (Strahler, A.N. 1969):

$$R_a = \frac{\bar{A}_u}{\bar{A}_u - 1}$$

where $\bar{A}_u$ is the mean area of given order of a basin.

When $\bar{A}_u = \frac{A_u}{N_u}$

where N_u = number of stream segments of a given order; and

A_u = the total area of all segments of the same order.

On the basis of the above formula mean basin areas (Table 2.10) have been obtained and thus with the help of $\bar{A}_u$, mean R_a of 20 sample basins have been derived (Table 2.11). The mean R_a of all the basins vary between 3.78 (Barar basin) and 16.81 (Kachchua basin). Likewise mean length ratio, it is expected that mean R_a vary among the basins of equal order but it should decrease comparatively in the basins of higher order. This rule holds good in several basins of the region.

Strahler's Law of Basin Area

Strahler has formulated the law of positive geometric progression of drainage basin area wherein "the mean basin areas of successive higher stream orders tend to form a geometric series beginning with mean area of the 1st order basin and increasing according to constant area ratio."

The above law is expressed by the following equation:

$$\bar{A}_u = \bar{A}_1 R_a^{(u-1)}$$

where $\bar{A}_1$ = the mean basin area of the 1st order; and

R_a = area ratio.

The correlation coefficient between orders and mean basin areas are obtained on the basis of the following regression equation:

$$\text{Log } y = \text{Log } a + bx$$

where y = the mean basin area;

x = order of the streams;

a = constant; and

b = regression coefficient.

Table 2.10: Mean basin areas ($\bar{A}u$) (in Sq. Kms.)

S. No.	*Drainage basins*	$\bar{A}u_1$	$\bar{A}u_2$	$\bar{A}u_3$	$\bar{A}u_4$	$\bar{A}u_5$
1.	Samrani	0.98	7.05	67.20	–	–
2.	Paroi	0.28	2.45	12.46	40.83	–
3.	Piprawal	0.40	3.71	17.78	40.38	–
4.	Barwa	0.10	0.58	3.63	22.35	45.13
5.	Ghinhal	0.24	1.47	6.53	46.83	–
6.	Jhuri	0.27	1.97	7.51	35.30	–
7.	Bagdara	0.32	2.10	11.04	52.25	–
8.	Kalibarah	0.20	1.20	4.64	17.80	40.38
9.	Maro	0.30	1.83	10.88	37.73	–
10.	Sukra	0.28	1.49	7.94	28.28	–
11.	Kulagara	0.23	1.50	7.73	7.90	–
12.	Pindrah	0.28	1.73	21.40	–	–
13.	Barar	0.18	0.97	4.26	13.54	29.30
14.	Kewai	0.29	1.71	8.00	23.19	77.35
15.	Ganta	0.21	1.15	4.87	16.40	–
16.	Jhurai	0.41	2.09	23.00	–	–
17.	Gauhia	0.63	8.20	32.10	–	–
18.	Kachchua	0.56	17.69	35.70	–	–
19.	Dhora	0.59	5.80	40.15	–	–
20.	Dingchi	0.51	2.74	8.89	40.25	–

Table 2.11: Area ratio (Ra)

S. No.	*Drainage basins*	Au_2/Au_1	Au_3/Au_2	Au_4/Au_3	Au_5/Au_4	R_a
1.	Samrani	7.19	9.53	–	–	8.36
2.	Paroi	8.75	5.09	3.28	–	5.71
3.	Piprawal	9.28	4.79	2.27	–	5.45
4.	Barwa	5.92	6.26	6.16	2.02	5.09
5.	Ghinhal	6.13	4.44	7.17	–	5.91
6.	Jhuri	7.30	3.81	4.70	–	5.27
7.	Bagdara	6.56	5.26	4.73	–	5.52
8.	Kalibarah	6.06	3.87	3.84	2.27	4.01
9.	Maro	6.14	5.95	3.47	–	5.19
10.	Sukra	5.32	5.33	3.62	–	4.76
11.	Kulagara	6.51	5.16	2.32	–	4.66
12.	Pindrah	6.18	12.37	–	–	9.28
13.	Barar	5.39	4.39	3.18	2.16	3.78
14.	Kewai	5.90	4.68	2.90	3.34	4.20
15.	Ganta	5.48	4.23	3.30	–	4.36
16.	Jhurai	5.10	11.00	–	–	8.05
17.	Gauhia	13.02	3.91	–	–	8.46
18.	Kachchua	31.59	2.02	–	–	16.81
19.	Dhora	9.83	6.92	–	–	8.38
20.	Dingchi	5.37	3.23	4.53	–	4.38

Regression lines (Fig. 2.9) are drawn on the basis of the above equation to test the linear relationship between drainage orders and mean basin areas. Out of 20 drainage basins, 4 basins (Jhurai, Samrani, Pindrah, and Dhora) of 3rd orders and 7 basins (Bagdara, Ganta, Maro, Dingchi, Ghinhal, Sukra and Jhuri) of 4th orders closely conform with the above law whereas the remaining basins show remarkable departure from the regression lines. In the case of all the 5th order basins, the above law does not hold good. The above relationship is also validated as the coefficient of correlation between these two variables range between +0.86 and +0.99.

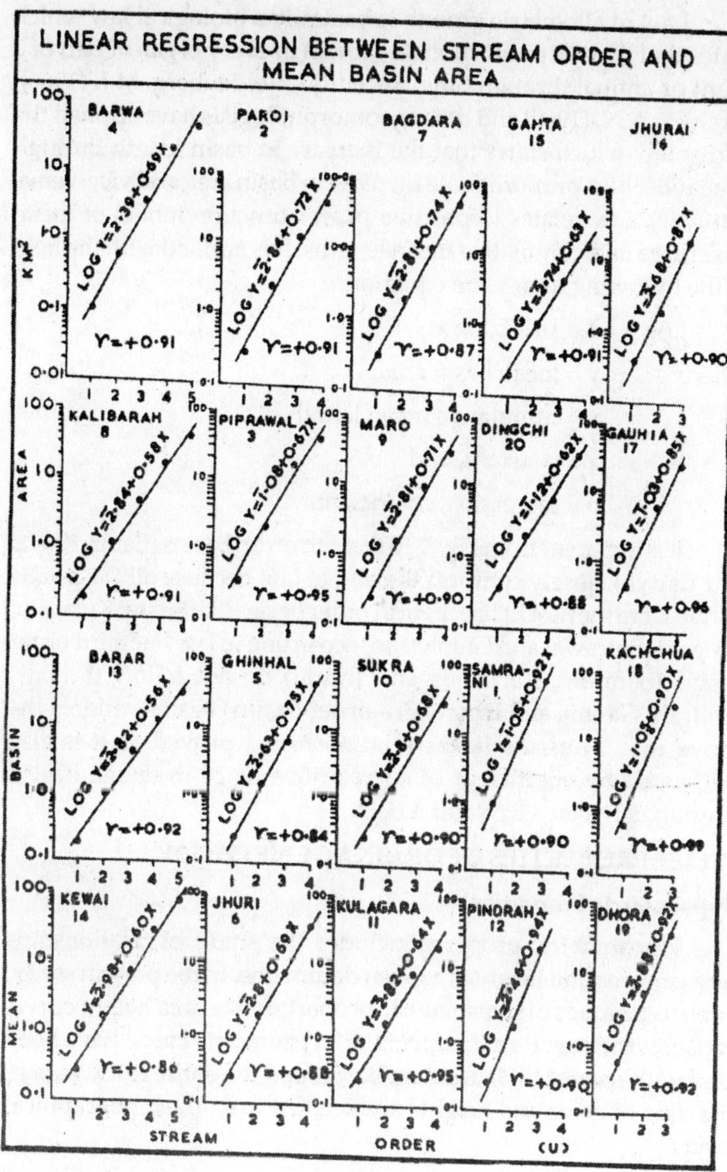
LINEAR REGRESSION BETWEEN STREAM ORDER AND MEAN BASIN AREA
MEAN BASIN AREA KM²
STREAM ORDER (U)
BARWA 4
γ=+0·91
PAROI 2
γ=+0·91
BAGDARA 7
γ=+0·87
GANTA 15
γ=+0·91
JHURAI 16
γ=+0·90
KALIBARAH 8
γ=+0·91
PIPRAWAL 3
γ=+0·95
MARO 9
γ=+0·90
DINGCHI 20
γ=+0·88
GAUHIA 17
γ=+0·96
BARAR 13
γ=+0·92
GHINHAL 5
γ=+0·84
SUKRA 10
γ=+0·90
SAMRA-NI 1
γ=+0·90
KACHCHUA 18
KEWAI 14
γ=+0·86
JHURI 6
γ=+0·88
KULAGARA 11
γ=+0·95
PINDRAH 12
γ=+0·90
DHORA 19
γ=+0·92

Fig. 2.9

Law of Allometric Growth

Law of allometric growth is basically a biological law which states that there is proportionate growth in every organ of part of a plant or animal through time. Recently, Woldenberg, M.J. (1966), Strahler, A.N. (1969) and other geomorphologists have applied the above law which states that the increase in basin length through time reflects in proportionate increase in basin area and vice-versa. Thus, the law relates to positive power function model of mean basin area and cumulative mean lengths. It is applied with the help of the following regression equation:

$$\text{Log } y = \text{Log } a + b \log x$$

where y = mean basin area,

x = cumulative mean length,

a = constant, and

b = regression, coefficient.

It is apparent from Fig. 2.10 that 5th order basins (Barar, Kewai and Barwa) closely conform the above law because all the points fall closer to the best fit line. Fourth order basins like Sukra, Kulagara, Dingchi, Piprawal and Ghinhal are according to law but third order basins (Samrani, Kachchua and Jhurai) do not follow the law. Pindrah, Gauhia and Dhora (3rd order basins) basins validate the above fact. Thus the linear relationship is proved as it is also evidenced by coefficient of correlations of 20 drainage basins ranging between + 0.88 and + 0.99.

RELIEF PROPERTIES OF DRAINAGE NETWORK

Hypsometric Properties

Hypsometric property includes the study of relationships between area and height at a given datum line. In the present study, three techniques of hypsometric properties like area height curve, hypsometric curve and percentage hypsometric curve have been used with respect to 20 drainage basins and the entire study region. The data of areas and heights are derived from the topographical sheets.

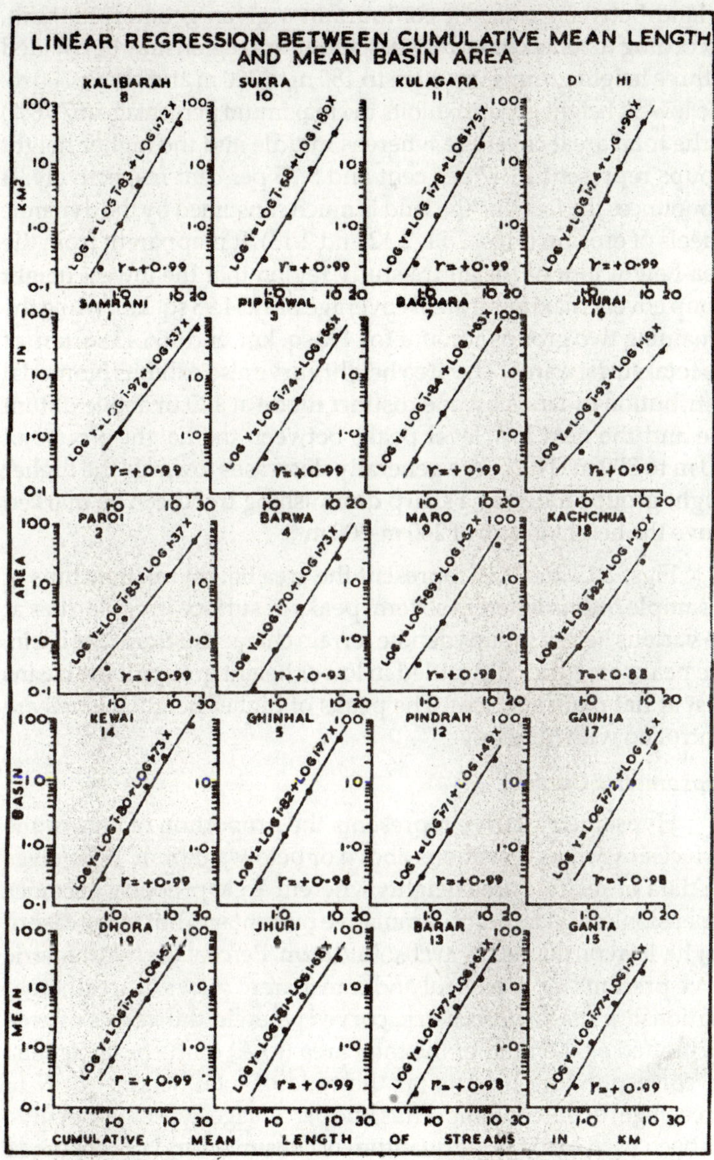
LINEAR REGRESSION BETWEEN CUMULATIVE MEAN LENGTH AND MEAN BASIN AREA
KALIBARAH 8
SUKRA 10
KULAGARA 11
D' HI
SAMRANI 1
PIPRAWAL 3
BAGDARA 7
JHURAI 16
PAROI 2
BARWA 4
MARO 9
KACHCHUA 18
KEWAI 14
GHINHAL 5
PINDRAH 12
GAUHIA 17
DHORA 19
JHURI 6
BARAR 13
GANTA 15
MEAN BASIN AREA IN KM2
CUMULATIVE MEAN LENGTH OF STREAMS IN KM

Fig. 2.10

Area-height Curve

The area height curve represents an absolute or relative areas of land between adjacent contours in a given geomorphic unit. According to relief characteristics, the entire region may be divided in three height groups viz. 60 m to 180 m to 300 m and above 300 m. The lowest height group exhibits the maximum percentage (67.78%) of the total areal coverage whereas middle and the higher height groups represent 26.47 per cent and 5.75 per cent respectively. It pronounces the fact that the land is much consumed by the dynamic wheels of erosion (Figs. 2.11, 2.12 and 2.13). It is apparent from the area-height plot of the entire study region that the lowest height group covers maximum areal coverage of 1254.98 sq. km. while the remaining two groups account for 490 sq. km. and 106.43 sq. km. of the total surface area. The area height curve also exhibits biomodal distribution of lands by one distinct mode at 120 m above datum line and the next low level peaks between the height groups of 260 m to 300 m. The curve generally decreases towards the higher height groups whereas a sharp diminishing trend can be marked above the height group of 260 m-300 m.

Figs. 2.12 and 2.13 represent the area height relationships of 18 sample basins wherein several peaks of surface irregularities at the various height groups denote terrain characteristics of the basin. The peaks are generally settled in lower height groups if the basins exist at flat plain surfaces. The peaks of higher height groups are controlled with lithology.

Hypsometric Curve

Hypsometric curve represents the proportion of area of the surface at various elevations above or below a datum. It involves the data of relative area heights wherein areal property becomes dimensionless (in terms of cumulative percentage) but in the case of height, it is considered as in absolute term. Percentage hypsometric curve presents more fruitful and transparent views of area-height relationship than hypsometric curves wherein the values of area are plotted as the ratio of the total area (a/A) of the basin against the corresponding heights (as the ratio of total heights (h/H in percentage) of the contours. Thus, the percentage hypsometric curve has been used to focus the quantum of consumed and unconsumed lands above datum line.

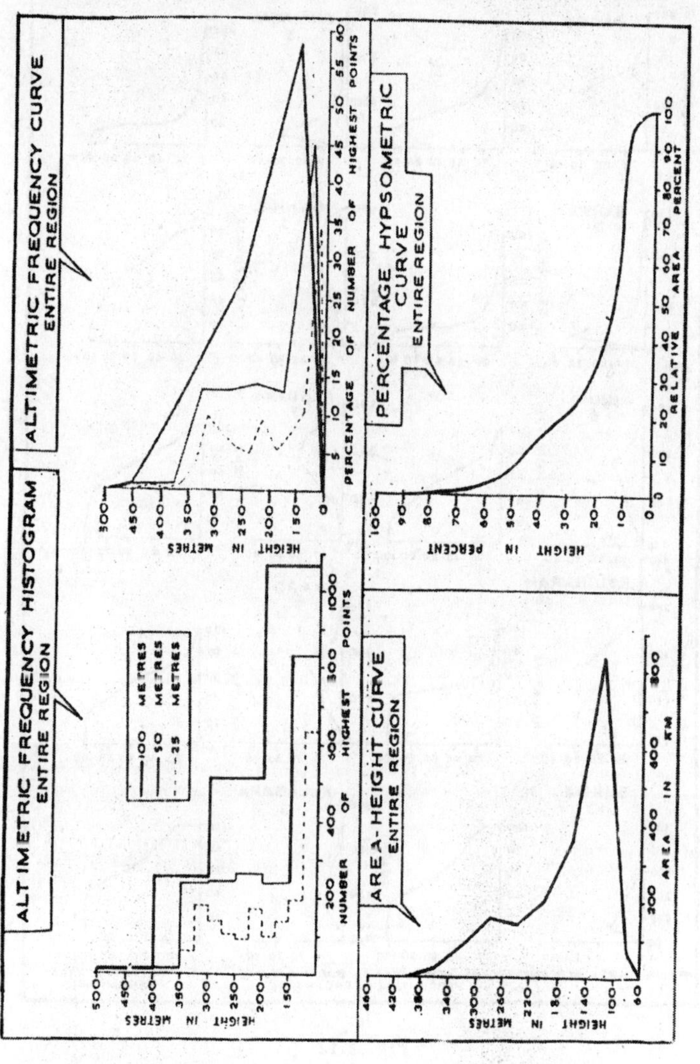
ALT IMETRIC FREQUENCY HISTOGRAM
ENTIRE REGION
100 METRES
50 METRES
25 METRES
HEIGHT IN METRES
NUMBER OF HIGHEST POINTS
ALTIMETRIC FREQUENCY CURVE
ENTIRE REGION
HEIGHT IN METRES
PERCENTAGE OF NUMBER OF HIGHEST POINTS
AREA-HEIGHT CURVE
ENTIRE REGION
HEIGHT IN METRES
AREA IN KM
PERCENTAGE HYPSOMETRIC CURVE
ENTIRE REGION
HEIGHT IN PERCENT
RELATIVE AREA PERCENT

Fig. 2.11

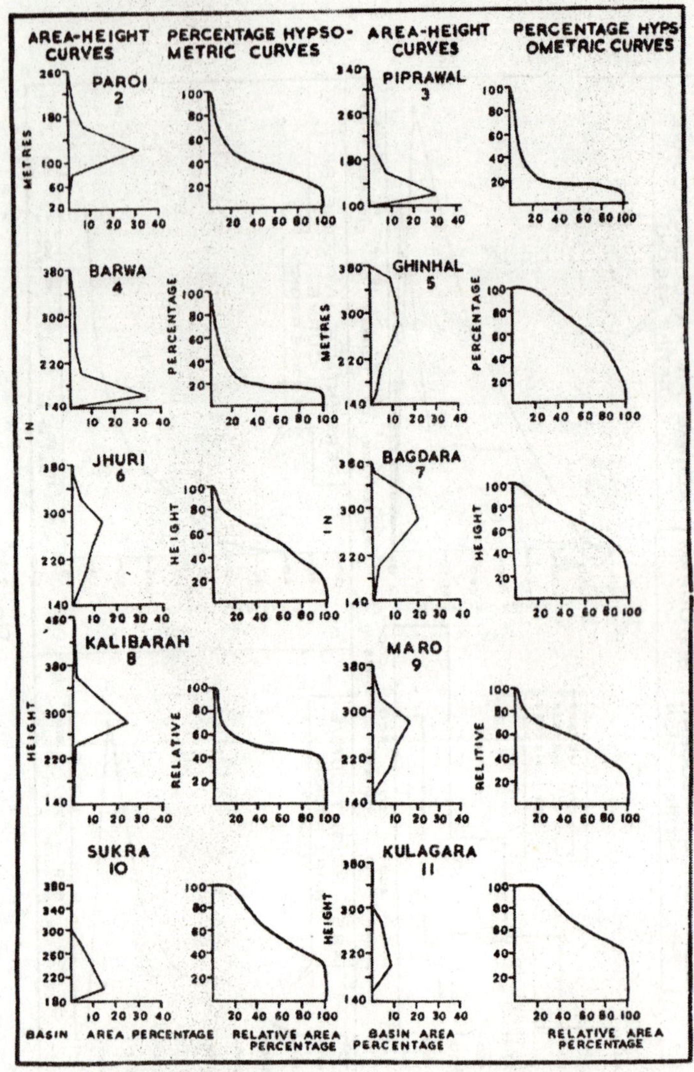
AREA-HEIGHT CURVES
PERCENTAGE HYPSOMETRIC CURVES
AREA-HEIGHT CURVES
PERCENTAGE HYPSOMETRIC CURVES
PAROI 2
PIPRAWAL 3
BARWA 4
GHINHAL 5
JHURI 6
BAGDARA 7
KALIBARAH 8
MARO 9
SUKRA 10
KULAGARA 11
METRES IN HEIGHT
PERCENTAGE HEIGHT RELATIVE
BASIN AREA PERCENTAGE
RELATIVE AREA PERCENTAGE
BASIN AREA PERCENTAGE
RELATIVE AREA PERCENTAGE

Fig. 2.12

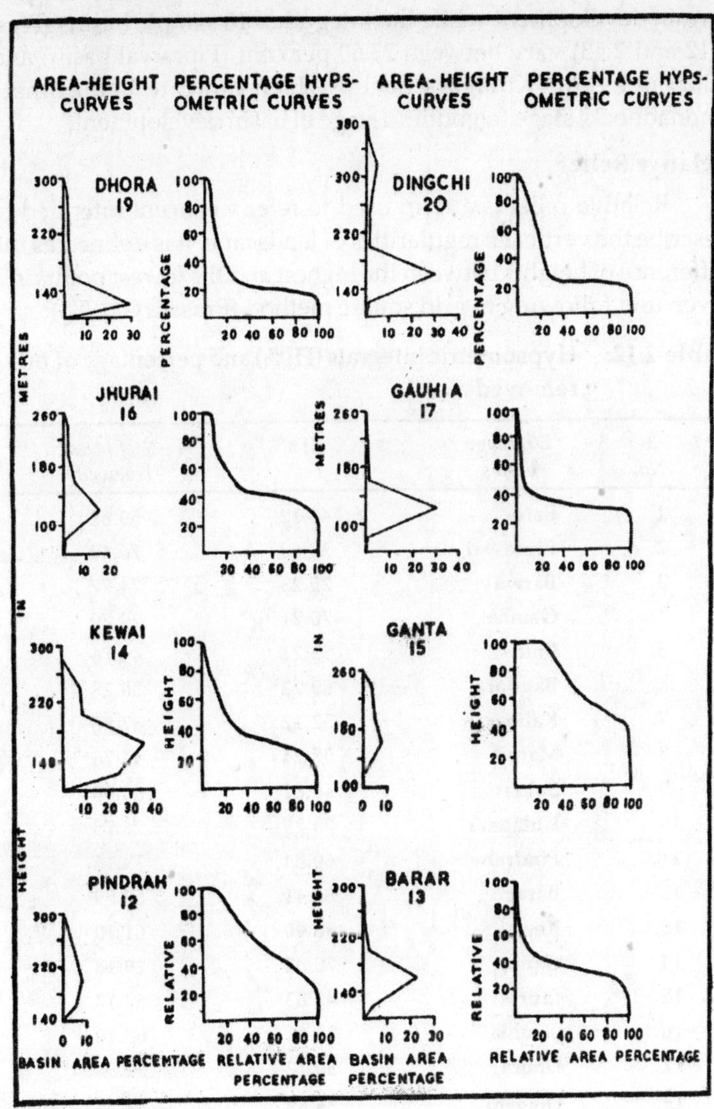
AREA-HEIGHT CURVES
PERCENTAGE HYPSOMETRIC CURVES
AREA-HEIGHT CURVES
PERCENTAGE HYPSOMETRIC CURVES
DHORA 19
DINGCHI 20
JHURAI 16
GAUHIA 17
KEWAI 14
GANTA 15
PINDRAH 12
BARAR 13
HEIGHT IN METRES
RELATIVE HEIGHT IN PERCENTAGE
BASIN AREA PERCENTAGE
RELATIVE AREA PERCENTAGE
BASIN AREA PERCENTAGE
RELATIVE AREA PERCENTAGE

Fig. 2.13

The hypsometric integral of the entire study region (Fig. 2.11, Table 2.12) accounts for 22.25 per cent indicates late or old stage of terrain development while the integral of 18 sample basins (Figs. 2.12 and 2.13) vary between 23.60 per cent (Piprawal basin) and 73.25 per cent (Kulagara basin) which indicate penultimate (monadnock) stage to youthful stage of basin development.

Relative Relief

Relative relief is a term used to refer a concept intended to describe the vertical irregularities of landscape. It is defined as the difference of heights between the highest and the lowest points in a given unit following to grid square method (Prasad, G. 1988).

Table 2.12: Hypsometric integrals (HI%) and percentage of mass removed

S. No.	*Drainage basins*	*HI (%)*	*% of mass removed*
1.	Paroi	40.12	59.88
2.	Piprawal	23.60	76.40
3.	Barwa	25.23	74.77
4.	Ghinhal	70.21	29.79
5.	Jhuri	55.24	44.76
6.	Bagdara	69.22	30.28
7.	Kalibarah	52.12	47.88
8.	Maro	55.24	44.76
9.	Sukra	65.62	34.38
10.	Kulagara	73.25	26.75
11.	Pindrah	60.84	39.16
12.	Barar	36.41	63.59
13.	Kewai	38.90	61.10
14.	Ganta	70.82	29.18
15.	Jhurai	47.83	52.17
16.	Gauhia	36.81	63.19
17.	Dhora	30.12	69.88
18.	Dingchi	36.83	63.17
	Entire region =	22.25	77.75

On the basis of the grid square method the relative relief has been obtained for basinal, intra-basinal and the region as a whole. The total frequency and frequency percentage under five categories have been also given in Tables 2.13 and 2.14 respectively.

The isopleth map (Fig. 2.14) of relative relief and Table 2.13 highlights highly asymmetrical distribution of frequencies in different categories. Extremely low, moderately low and low relative relief categories together account for 74.90 per cent of the total frequencies whereas only 17.43 per cent of the total frequencies are registered in moderate categories. The minimum (7.67%) concentrations are marked in moderately high relative relief category. The intra-basinal percentage are also too much closer to regional explanation. Table 2.14 showing frequency distribution of relative relief for 20 sample basins reveals the fact that low moderate and moderately high relative relief categories register 70.08 per cent, 22.21 per cent and 7.71 per cent of the total frequencies respectively. The above result also coincides the regional analysis. But there are large variations from one basin to another. Samrani and Kachchua basins show their total frequencies in low relative relief category and indicate flat land terrain. It is interesting that out of 20, 16 drainage basins occur more than 60 per cent of their total frequencies in low relative relief categories ranging between 5.56 per cent (Ganta basin) and 100 per cent (Samrani and Kachchua basins).

The moderate relative relief category ranges between 3.23 per cent and 88.89 per cent (Ganta basin). Maximum concentration is noticed in Ganta (88.89%), Jhuri (68.50%) and Ghinhal (51.22%) basins. Kalibarah basin represents an ideal distribution of frequencies in all the categories. Similar results are obtained in Paroi, Bagdara and Dingchi basins.

The isopleth map (Fig. 2.14) reveals that the area north of 25° 10'N latitude is covered by low relative relief which is more than 90 per cent of the total area. Several monadnock of northern alluvial plain are dotted with moderate and moderately high relative relief. Most of the top surfaces and valley floor basement of southern upland are well confined with low relative relief due to flattish terrain and low degree of dissection.

Table 2.13: Frequency distribution of relative relief (R_R)

Sl. No.	Areal unit	Relative relief categories (in m)									
		Below-15 Extremely low		15-30 Moderately low		30-60 Low		60-120 Moderate		Above 120 Moderately high	
		F	%	F	%	F	%	F	%	F	%
1.	Entire region	914	49.40	223	12.10	248	13.40	323	17.43	142	07.67
2.	Intra-basinal (20 selected basins)	290	37.85	76	14.08	137	18.15	172	22.21	65	07.71

Table 2.14: Frequency distribution of relative relief (sample drainage basins)

Sl. No.	Drainage basins	Relative relief categories										Total frequencies
		Below-15 R_{EL}		15-30 R_{ML}		30-60 R_L		60-120 R_M		Above 120 R_{MH}		
		F	%	F	%	F	%	F	%	F	%	
1	2	3	4	5	6	7	8	9	10	11	12	13
1.	Samrani	67	100.00	–	–	–	–	–	–	–	–	67
2.	Paroi	25	55.56	03	06.67	11	24.44	06	13.33	–	–	45
3.	Piprawal	27	71.05	01	02.63	01	02.63	04	10.53	05	13.16	38
4.	Barwa	11	21.74	05	10.87	03	06.52	07	15.22	20	45.65	46
5.	Ghinhal	02	04.88	09	21.95	02	04.88	21	51.22	07	17.07	41
6.	Jhuri	–	–	–	–	08	22.86	24	68.57	03	08.57	35
7.	Bagdara	05	09.43	05	09.43	24	45.28	18	33.96	01	01.90	53
8.	Kalibarah	10	24.39	10	24.39	10	24.39	05	12.20	06	14.63	41
9.	Maro	04	11.11	05	13.89	17	47.72	10	27.78	–	–	36
10.	Sukra	04	13.80	05	17.24	11	37.93	09	31.03	–	–	29
11.	Kulagara	01	05.56	05	27.78	07	38.88	05	27.78	–	–	18
12.	Pindrah	02	09.09	07	31.82	08	36.36	05	22.73	–	–	22

(Contd...)

1	2	3	4	5	6	7	8	9	10	11	12	13
13.	Barar	04	12.90	15	51.61	10	32.26	02	03.23	–	–	31
14.	Kewai	13	17.33	24	32.00	20	26.67	18	24.00	–	–	75
15.	Ganta	–	–	01	05.56	–	–	16	88.89	00	05.56	18
16.	Jhurai	14	58.33	03	12.50	–	–	07	29.17	–	–	24
17.	Gauhia	22	66.67	05	15.15	–	–	03	09.09	03	09.09	33
18.	Kachchua	37	100.00	–	–	–	–	–	–	–	–	37
19.	Dhora	24	60.00	–	–	02	05.00	06	15.00	08	20.00	40
20.	Dingchi	18	42.86	03	07.14	03	07.14	06	14.29	12	28.57	42
Total frequency and average %		290	37.85	106	14.08	137	18.15	172	24.90	65	08.21	771

Average % = R_L = 70.08, R_M = 22.21%, R_{MH} = 7.71%

R_{EL} = Extremely low relative relief, R_L = Low relative relief, R_{MH} = Moderately high relative relief, R_{ML} = Moderately low relative relief, R_M = Moderate relative relief.

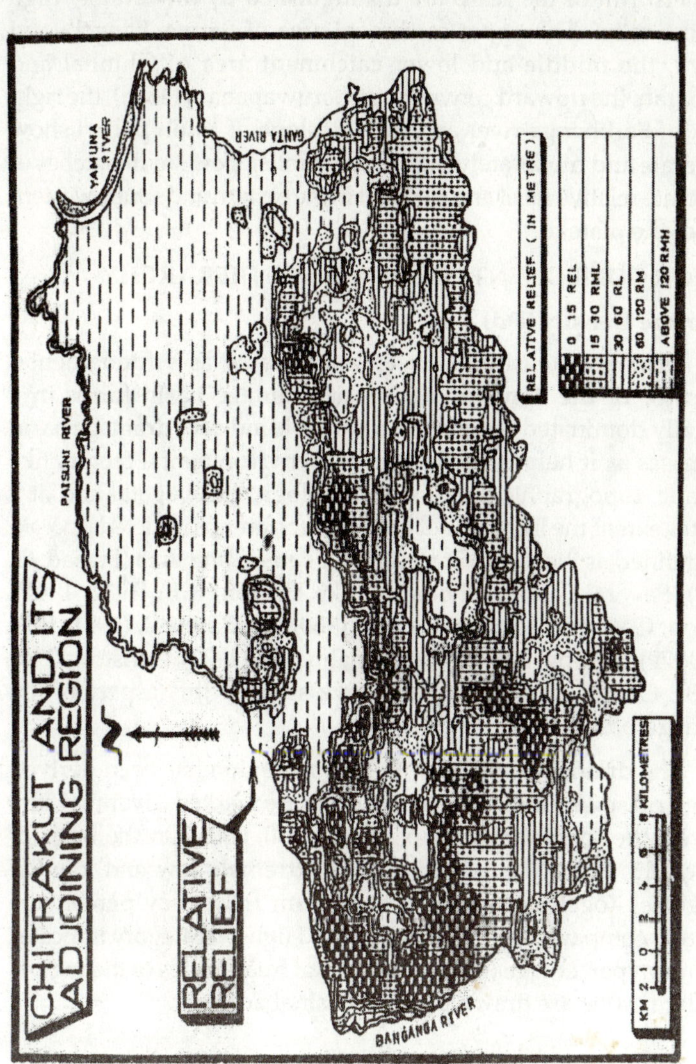
CHITRAKUT AND ITS ADJOINING REGION
RELATIVE RELIEF
YAMUNA RIVER
GANTA RIVER
PAISUNI RIVER
BANGANGA RIVER
RELATIVE RELIEF (IN METRE)
0 15 REL
15 30 RML
30 60 RL
60 120 RM
ABOVE 120 RMH
KM 2 1 0 2 4 6 8 KILOMETRES

Fig. 2.14

Relative relief increases gradually from the valley floor basement to the ridges. The western fringes of the plateau and the northern rim of the scarp are distinguished by moderately high relative relief. The upper catchment area of Barwa, Dingchi and Dhora, the middle and lower catchment area of Ghinhal and Kalibarah, the upward convexity of Geruwapahar (446 m), the right flank of Sarbhanga river and the sideslope of Knoh pahari show moderate and moderately high relative relief. Few small patches of moderate relative relief are observed in the eastern and south-western part of the plateau.

MAJOR COMPONENTS OF DRAINAGE HIERARCHY

Drainage Density (Dd)

Drainage density has been appreciated as a fundamental indicator of the dynamic nature of drainage basin forms in a fluvially dominated terrain. It is a valuable index of drainage basin processes as it helps to reflect several environmental factors like climatic, topographic, lithologic, pedologic and vegetal and at a greater extent the influence of human interference inflicted upon it. It is defined as the length of stream per square unit area (Prasad, G. 1987). Several Geomorphologists like Horton, R.E. (1932, 1945), Carton, C.W. and Langbein, W.B. (1960), Gardiner, V. (1971, 1974 and 1979), McCoy, R.M. (1970), Donahue, J.J. (1972), Cariston, C.W. (1963), Catton, C.A. (1964) etc. have studied this aspect of the drainage basins.

The drainage density of the whole region has been derived using grid square method wherein grids are marked covering 1 km^2 area on the topographic maps (Scale 1:50, 000). On the basis of Table 2.15, it may be explained that extremely low and low Dd categories together represent maximum frequency percentage (58.98%) comparatively to other groups. High Dd category indicates minimum percentage (0.42%) of the total frequencies of the region. Similar results are drawn by intra-basinal analysis.

Table 2.15: Frequency distribution of drainage density (Dd) (whole region) and intra-basinal

Dd categories	*Frequencies*		*Percentage*	
	Regional	*Infra-basinal*	*Regional*	*Infra-basinal*
Below 1 (Dd_{EL})	549	139	29.68	18.03
1-2 (Dd_L)	542	256	29.30	33.20
2-4 (Dd_M)	687	338	37.14	43.84
4-6 (Dd_{MH})	64	36	3.46	4.67
Above 6 (Dd_H)	8	02	0.42	0.26
Total	1850	771	100.00	100.00

Dd_{EL} = Extremely low drainage density;
Dd_L = Low drainage density;
Dd_M = Moderate drainage density;
Dd_{MH} = Moderately high drainage density; and
Dd_H = High drainage density.

The frequency distribution (Table 2.16) in 20 drainage basins show slight variations. It is apparent that the basins which are developed over northern plain (specially Samrani and Kachchua basins) or partly on plateau and partly on plain like Piprawal, Jhurai, Gauhia, Dhora and Dingchi are characterized by low and moderate Dd categories wherein low Dd category is much dominating. Basins (Ghinhal, Jhuri, Bagdara, Kalibarah, Maro, Sukra, Kulagara, Pindrah, Barar and Ganta) of highly dissected terrain carry low to moderately high and high drainage density.

Fig. 2.15 illustrates a view of high and moderately high drainage density region in a few small patches in the lower reaches of Barwa basin, along the downslope of several dissected hills, along the upper reaches of rivers Ganta, Sukra, Jhuri, Kalibarah and Burranala and some other highly dissected areas.

The moderate drainage density region can be marked in the south of 25° 10′ N latitude. The upper reaches of Samrani, middle and lower reaches of Ohan and lower sections of Geduwa and Kachchua rivers are well confined with moderate drainage density.

Table 2.16: Drainage density (Dd) categories (frequency distribution of sample basins)

Sl. No.	Drainage basins	Drainage density (Dd) categories										Total frequencies
		Below-1 Dd_{EL}		1-2 Dd_{L}		2-4 Dd_{M}		4-6 Dd_{MH}		Above 6 Dd_{M}		
		F	%	F	%	F	%	F	%	F	%	
1	2	3	4	5	6	7	8	9	10	11	12	13
1.	Samrani	27	40.30	26	38.80	14	20.90	–	–	–	–	67
2.	Paroi	12	26.67	19	42.22	14	31.11	–	–	–	–	45
3.	Piprawal	09	23.68	16	42.11	13	34.21	–	–	–	–	38
4.	Barwa	05	10.87	13	28.26	13	28.26	13	28.26	02	04.35	46
5.	Ghinhal	03	07.32	07	17.07	24	58.54	07	17.07	–	–	41
6.	Jhuri	01	02.86	08	22.86	24	68.57	02	05.71	–	–	35
7.	Bagdara	07	13.21	22	41.51	24	45.18	–	–	–	–	53
8.	Kalibarah	02	04.88	10	24.39	26	63.41	03	07.32	–	–	41
9.	Maro	03	08.33	08	22.22	25	69.45	–	–	–	–	36
10.	Sukra	01	03.45	11	37.93	15	51.72	02	06.90	–	–	29
11.	Kulagara	–	–	06	33.33	12	66.67	–	–	–	–	18
12.	Pindrah	–	–	06	27.27	16	72.73	–	–	–	–	22

(Contd...)

1	2	3	4	5	6	7	8	9	10	11	12	13
13.	Barar	02	06.45	09	20.03	20	65.52	–	–	–	–	31
14.	Kewai	07	09.33	15	20.00	48	64.00	05	06.67	–	–	75
15.	Ganta	02	11.11	02	11.11	10	55.56	04	22.22	–	–	18
16.	Jhurai	05	20.83	11	45.83	08	33.34	–	–	–	–	24
17.	Gauhia	10	30.30	14	42.43	09	27.27	–	–	–	–	33
18.	Kachchua	22	59.46	12	32.43	03	08.11	–	–	–	–	37
19.	Dhora	13	32.50	19	47.50	08	20.00	–	–	–	–	40
20.	Dingchi	08	19.05	22	52.38	12	28.57	–	–	–	–	42
Total frequency and average %		139	18.03	256	33.20	338	43.84	36	04.67	02	00.26	771

Dd_L = 51.23% Dd_M = 43.84%

Dd_{MH} = 4.67% Dd_H = 0.26%

Dd_{EL} = Extremely low drainage density Dd_{MH} = Moderately high drainage density

Dd_L = Low drainage density Dd_H = High drainage density

Dd_M = Moderate drainage density

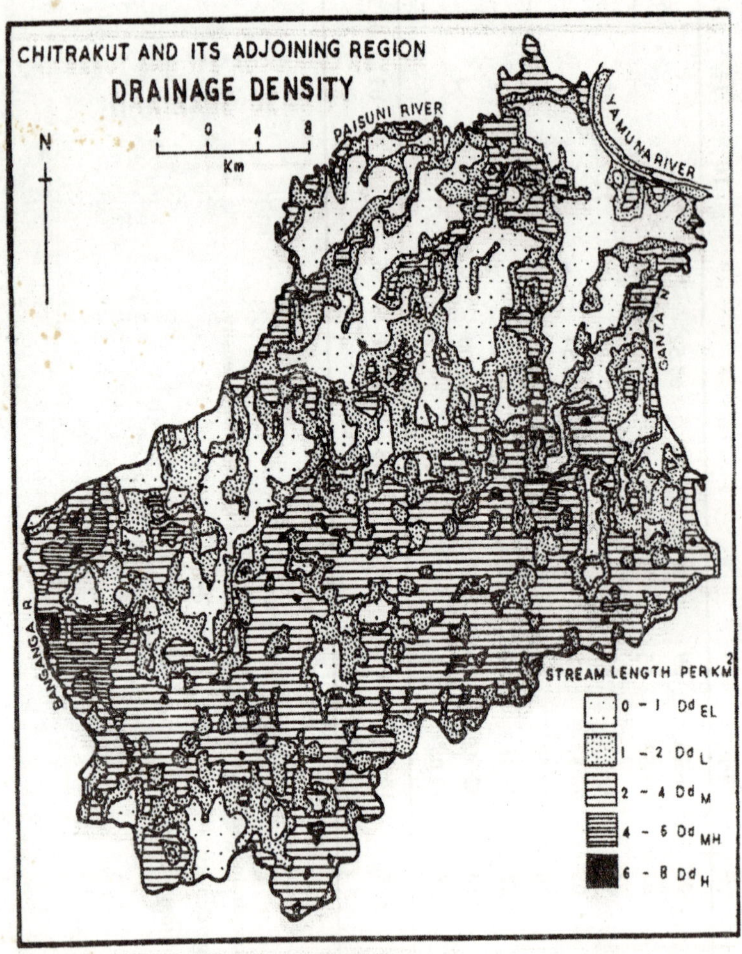
CHITRAKUT AND ITS ADJOINING REGION
DRAINAGE DENSITY
PAISUNI RIVER
YAMUNA RIVER
GANTA N.
BANGANGA R.
N
4 0 4 8
Km
STREAM LENGTH PER KM2
0 - 1 Dd EL
1 - 2 Dd L
2 - 4 Dd M
4 - 6 Dd MH
6 - 8 Dd H

Fig. 2.15

Low drainage density is dominating over the extensive area of Yamuna plain between 25° 10′ N latitude and 25° 25′ N latitude. It occupies about 60 per cent of the total surface area. Few scattered patches are also seen over the hill tops, free-faces of the scarps and ridges and in the lower catchment area of the basins.

Hydrological Concept of Drainage Density

The mean drainage density (d) for 20 drainage basins has been calculated according to orders and shown in Table 2.17 and Fig. 2.16. It is apparent that mean drainage density (a) has no definite trends according to drainage orders. Basins like Samrani, Gauhia, Dhora and Barar indicate decreasing trends because of flat plain country. Bagdara, Piprawal, Dingchi, Kalibarah and Ganta show sudden swift in their 4th order due to maximum stream lengths caused by slope steepness. Jhurai basin represents abnormal distribution whereas Ghinhal exhibits opposite trends. More or less Pindrah basin depicts an ideal distribution. On the basis of the above discussion it may be concluded that the drainage density generally decreases according to orders if the structural stability is steady state in nature. If all the environmental factors are equilibrium in nature, the drainage density will tend to be constant. The instability of natural factors cause the high drainage density. The terminal points between the two contrasting terrain and steep slope sides produce higher drainage density. Accelerated soil erosion on bare rock surfaces due to gullying and rilling causes maximum stream segments and the higher drainage density (Prasad, G. 1987).

Drainage Density Ratio (β)

Drainage density ratio is the ratio of drainage density of streams of a given order to the density of streams of the next higher order. The drainage density of streams of different orders in a given drainage basin tend closely to approximate an inverse geometric series in which the first term is unity and the ratio is the drainage density ratio (Prasad, G. 1987, Jayaswal, S.N.P. and Prasad, G. 1989).

Table 2.17: Measures of drainage density and drainage density ratios in different drainage basins

Sl. No.	Drainage basins	Drainage density (a) of given orders					Drainage density ratio (β)				
		α^1	α^2	α^3	α^4	α^5	$\beta_1 = \frac{\alpha^1}{\alpha^2}$	$\beta_2 = \frac{\alpha^2}{\alpha^3}$	$\beta_3 = \frac{\alpha^3}{\alpha^4}$	$\beta_4 = \frac{\alpha^4}{\alpha^5}$	
1.	Samrani	1.44	1.04	0.92	–	–	1.38	1.13	–	–	1.26
2.	Paroi	2.81	0.88	1.68	0.93	–	3.19	0.52	1.81	–	1.84
3.	Piprawal	2.21	1.18	0.91	1.42	–	1.79	1.30	0.64	–	1.24
4.	Barwa	4.60	2.61	2.03	1.44	2.00	1.76	1.29	1.41	0.72	1.30
5.	Ghinhal	2.38	2.30	2.59	6.49	–	1.03	0.89	0.40	–	0.77
6.	Jhuri	2.75	1.97	2.51	1.87	–	1.40	0.78	1.34	–	1.17
7.	Bagdara	2.26	1.58	0.99	1.57	–	1.43	1.60	0.63	–	1.22
8.	Kalibarah	2.89	2.33	2.22	2.42	1.36	1.24	1.05	0.92	1.78	1.25
9.	Maro	2.42	1.91	2.03	2.19	–	1.27	0.94	0.93	–	1.05
10.	Sukra	2.69	2.05	2.56	1.15	–	1.31	0.80	2.23	–	1.45
11.	Kulagara	2.85	1.86	2.15	1.88	–	1.53	0.87	1.14	–	1.18
12.	Pindrah	2.56	2.36	2.45	–	–	1.08	0.96	–	–	1.02
13.	Barar	3.11	2.10	1.86	1.84	1.60	1.48	1.13	1.01	1.15	1.19
14.	Kewai	2.42	2.18	2.86	1.94	1.41	1.11	0.76	1.47	1.38	1.18
15.	Ganta	3.03	2.78	2.73	2.99	–	1.09	1.02	0.91	–	1.01
16.	Jhurai	2.37	3.67	1.26	–	–	0.65	2.91	–	–	1.78
17.	Gauhia	1.72	1.15	0.97	–	–	1.50	1.19	–	–	1.35
18.	Kachchua	1.38	0.68	1.54	–	–	2.03	0.44	–	–	1.24
19.	Dhora	1.72	1.03	0.96	–	–	1.67	1.07	–	–	1.37
20.	Dingchi	1.86	1.27	1.01	1.22	–	1.46	1.26	0.83	–	1.18

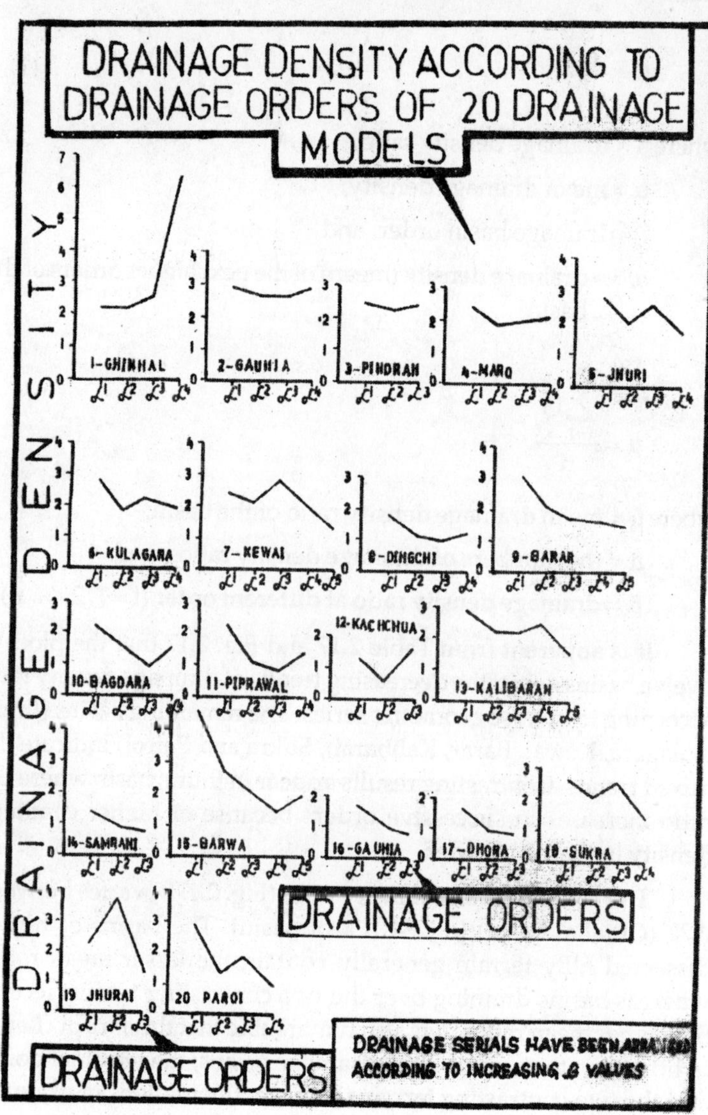
DRAINAGE DENSITY ACCORDING TO DRAINAGE ORDERS OF 20 DRAINAGE MODELS
DRAINAGE DENSITY
1-GHINHAL
2-GAUHIA
3-PINDRAH
4-MARO
5-JHURI
6-KULAGARA
7-KEWAI
8-DINGCHI
9-BARAR
10-BAGDARA
11-PIPRAWAL
12-KACHCHUA
13-KALIBARAN
14-SAMRANI
15-BARWA
16-GAUHIA
17-DHORA
18-SUKRA
19 JHURAI
20 PAROI
DRAINAGE ORDERS
DRAINAGE ORDERS
DRAINAGE SERIALS HAVE BEEN ARRANGED ACCORDING TO INCREASING ß VALUES

Fig. 2.16

On the basis of the Hortonian law, the present investigator has proposed a new formula for the calculation of drainage density ratio:

$$\beta = \frac{\alpha^{i}}{\alpha^{i+1}} \quad \ldots (1)$$

where β = drainage density ratio;

α = mean drainage density;

i = drainage basin order; and

α^{i+1} = drainage density (mean) of the next higher orders of the basin

$$\mu = \frac{\sum_{i=1}^{n} \beta_i}{n}$$

where μ = mean drainage density ratio of the basin;

n = the numbers of drainage density ratios; and

β_1= drainage density ratio at different order (i = 1, 2, ... n).

It is apparent from Table 2.17 and Fig. 2.17 that the plots of twelve basins depict the decreasing trends of drainage density ratio according to inverse geometric series. The remaining basins (Jhuri, Kulagara, Kewai, Barar, Kalibarah, Sukra and Pairoi) indicate the mixed trends. Contrasting results appear in Jhurai basin where the ratio increases in successive orders because of higher drainage density in their 2nd order.

The mean drainage density ratio (Fig. 2.17B) varies between 0.77 (Ghinhal basin) to 1.84 (Paroi basin). The basins of highly dissected hilly terrain generally contain the lesser mean ratios whereas basins draining over the two contrasting terrain receive the higher mean ratios. It is also remarkable that the abrupt change in lithology, the area of small drainage basins, the terminal points and the two contrasting terrains indicate fluctuations in drainage density ratio.

Drainage Texture (Dt)

Drainage texture is a fundamental, geomorphic concept and an important morphometric variable of network analysis which elucidates the nature of relative spacing of drainage lines. It was discussed by Horton, R.E. (1945), Smith, K.G. (1950), Singh, S. (1976), Prasad, G. (1987) etc. The author has advocated the cooked opinion of several geomorphologists that Dt must be used for the determination of relative spacing of the streams in a unit area along a linear direction and proposed a general formula for the deviation of Dt in the following manner:

$$Dt = As = \frac{1}{(x+y)/2} = \frac{2}{x+y}$$ = average spacing covered by one crossing ...(1)

where Dt (As) = drainage texture which is the average spacing of one crossing on the area of 1 km^2

$x = \frac{C_1 + C_2 + C_3 + C_4}{4}$ = average crossing per unit length on four edges of grid square of 1 km^2

where $C_1 \longrightarrow C_4$ = number of intersections between streams and four grid edges.

while $y = \frac{N}{\sqrt{2\Pi}}$ = average crossing per unit length on the circumference of the given circle of an area of ½ km^2

$$= \frac{\text{Number of crossings}}{\text{Length of circumstance}}$$

Where N = numbers of stream crossings intersecting the circumstance of a circle of area ½ km^2 drawn in the centre of each grid square of 1 km^2

Sometime, it may happen that the shorter length of streams specially in limestone area may not cross the circumference of a circle of area ½ Km^2 drawn in the centre of grid square of 1 km^2. Therefore, it is necessary again to draw the second circle of an area occupying ¼ km^2 within the first circle. If it is so, the above formula may be considered in the following manner:

$$Dt = As = \frac{3}{(x+y+z)} \qquad ...(2)$$

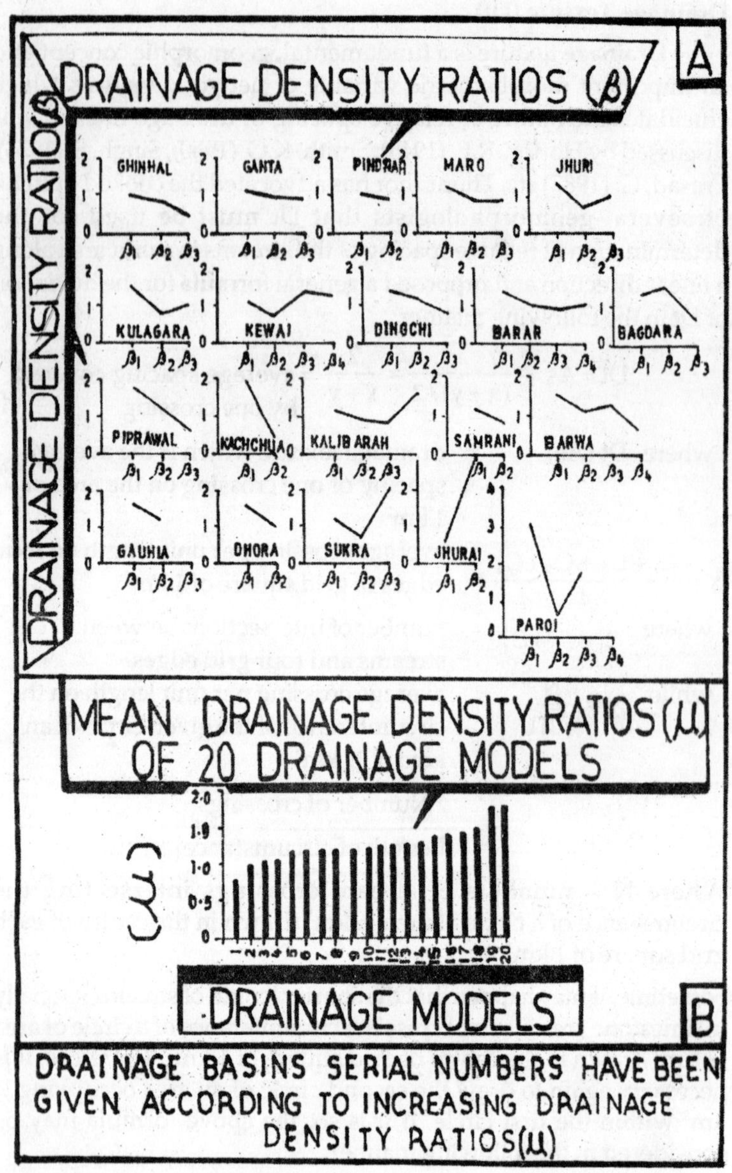
DRAINAGE DENSITY RATIOS (β)
A
DRAINAGE DENSITY RATIOS (β)
GHINHAL
GANTA
PINDRAH
MARO
JHURI
KULAGARA
KEWAI
DINGCHI
BARAR
BAGDARA
PIPRAWAL
KACHCHUA
KALIBARAH
SAMRANI
BARWA
GAUHIA
DHORA
SUKRA
JHURAI
PAROI
MEAN DRAINAGE DENSITY RATIOS (μ) OF 20 DRAINAGE MODELS
(μ)
DRAINAGE MODELS
B
DRAINAGE BASINS SERIAL NUMBERS HAVE BEEN GIVEN ACCORDING TO INCREASING DRAINAGE DENSITY RATIOS (μ)

Fig. 2.17

where x and y are as elaborated in equation 1

while $z = \frac{N}{\sqrt{\Pi}}$

where N = Number of stream crossings intersecting the circumference of a circle of area ¼ km^2 drawn within the first circle.

On the basis of the above formula, the Dt of the region has been calculated and the frequency percentage of the derived values are given in Table 2.18 and 2.19. Table 2.18 denotes that the coarse texture is much dominating in the region. The intra-basinal analysis is also similar in character.

The isopleth map (Fig. 2.18) deals with the spatial arrangement of five Dt categories. On the basis of its nature and characteristics it may be classified in the following Dt regions:

1. *Fine Dt Region:* Specially in the western fringe of the upland.
2. *Moderate Dt Region:* Generally the southern part of the plateau along the toeslope of hills.
3. *Coarse Dt Region:* Northern alluvial country between 25° 10′ N latitude and 25° 27′ N latitude.

Table 2.18: Frequency distribution of drainage texture (Dt) (Regional and Intrabasinal)

Dt categories	*Frequencies*		*Percentage*	
	Regional	*Intra bacinal*	*Regional*	*Intra-basinal*
Below 0.2 (T_{VF})	04	02	0.22	0.26
0.2-0.4 (T_F)	167	92	9.03	11.93
0.4-0.6 (T_M)	415	208	22.53	26.99
0.6-0.8 (T_C)	336	149	18.16	19.31
Above (T_{VC})	928	320	50.16	41.51
Total	1850	771	100.00	100.00

T_{VF} = Very fine drainage texture;
T_F = Fine drainage texture;
T_M = Moderate drainage texture;
T_C = Coarse drainage texture; and
T_{VC} = Very coarse drainage texture.

Table 2.19: Drainage texture (Dt) categories (Frequency distribution of sample basins)

Sl. No.	Drainage basins	Drainage texture categories										Total frequencies
		Below-0.2		0.2-0.4		0.4-0.6		0.6-0.8		Above 0.8		
		F	%	F	%	F	%	F	%	F	%	
1	2	3	4	5	6	7	8	9	10	11	12	13
1.	Samrani	–	–	–	–	11	16.42	09	13.43	47	70.15	67
2.	Paroi	–	–	02	04.44	10	22.22	07	15.56	26	57.78	45
3.	Piprawal	–	–	01	02.63	06	15.79	10	26.32	21	55.26	38
4.	Barwa	01	02.17	20	43.48	02	04.35	07	15.22	16	34.78	46
5.	Ghinhal	–	–	08	19.51	17	41.47	08	19.51	08	19.51	41
6.	Jhuri	–	–	05	14.29	16	45.71	09	25.71	05	14.29	35
7.	Bagdara	–	–	–	–	12	22.64	13	24.53	28	52.83	53
8.	Kalibarah	–	–	09	21.95	11	26.83	14	34.15	07	17.07	41
9.	Maro	–	–	01	02.78	23	63.89	04	11.11	08	22.22	36
10.	Sukra	–	–	02	06.90	14	48.28	08	27.58	05	17.24	29
11.	Kulgara	–	–	05	27.78	06	33.33	03	16.67	04	22.22	18
12.	Pindrah	–	–	02	09.09	11	50.00	05	22.73	04	18.18	22
13.	Barar	–	–	07	22.58	12	38.72	06	19.35	06	19.35	31

(Contd...)

1	2	3	4	5	6	7	8	9	10	11	12	13
14.	Kewai	–	–	16	21.33	30	40.00	14	18.67	15	20.00	75
15.	Ganta	–	–	10	55.56	03	16.67	02	11.11	03	16.16	18
16.	Jhurai	–	–	–	–	05	20.83	05	20.83	14	58.34	24
17.	Gauhia	01	03.03	–	–	06	18.18	06	18.18	20	60.61	33
18.	Kachchua	–	–	–	–	02	05.41	03	08.11	32	86.48	37
19.	Dhora	–	–	02	05.00	03	07.50	08	20.00	27	67.50	40
20.	Dingchi	–	–	02	04.76	08	19.05	08	19.05	24	57.14	42
Total frequency and average frequency %age		02	00.26	92	11.93	208	26.98	149	19.33	320	41.50	771

Fine drainage texture (T_F) = 12.19%   Moderate drainage texture (T_M) = 26.98%

Coarse drainage texture (T_C) = 60.83%

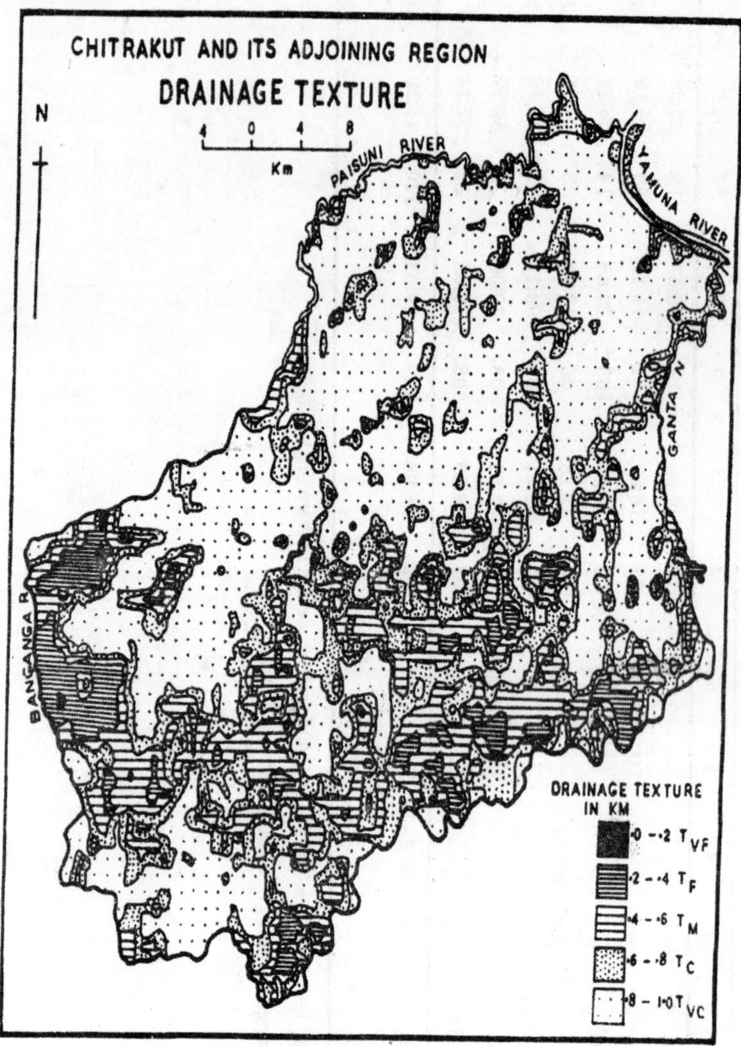
CHITRAKUT AND ITS ADJOINING REGION
DRAINAGE TEXTURE
N
4 0 4 8
Km
PAISUNI RIVER
YAMUNA RIVER
GANTA N
BANGANGA R
DRAINAGE TEXTURE IN KM
0 – ·2 TVF
·2 – ·4 TF
·4 – ·6 TM
·6 – ·8 TC
·8 – 1·0 TVC

Fig. 2.18

Stream Frequency (S_F)

Stream frequency or drainage frequency refers as the total number of drainage lines in a drainage basin or a morphounit by the corresponding area of the said unit. It was discussed by Horton, R.E. (1932 and 1945) and Strahler, A.N. (1957) but the proposal of the above scholars are not applicable as an easy process.

On the basis of the total number of drainage lines in a one km^2 grid area, the stream frequency of all the drainage basins and the entire study region has been derived. The derived values are classified in five categories wherein the frequency percentage are also taken into consideration (Tables 2.20 and 2.21).

Table 2.20: Frequency distribution of stream frequency (S_F) (Regional and Intrabasinal)

S_F categories	*Frequencies*		*Percentage*	
	Regional	*Intra-basinal*	*Regional*	*Intra-basinal*
Below 2 (F_{VP})	547	146	29.56	18.92
2-5 (F_P)	751	323	40.60	41.92
5-10 (F_M)	458	250	24.76	32.42
10-20 (F_{MH})	90	51	4.86	6.61
Above 20 (F_H)	04	01	0.22	0.13
Total	1850	771	100.00	100.00

F_{VP} = Very poor stream frequency;
F_P = Poor stream frequency;
F_M = Moderate stream frequency;
F_{MH} = Moderately high stream frequency; and
F_H = High stream frequency

According to Table 2.20, it is clear that the poor stream frequency groups cover maximum frequency percentage (70.16% at regional and 60.84% of intra-basinal level). It is 24.76 per cent in regional analysis and 32.42 per cent in intra-basinal study under moderate stream frequency category. The moderately high and high stream frequency percentage are minimum. Table 2.21 also indicates the domination of poor S_F categories in all the sample basins. The isopleth map (Fig. 2.19) shows five regions of S_F categories. It is remarkable that the smaller area is shown by high S_F coverage while the larger area (in the north of 25° 10′ N latitude) is covered with low (poor) stream frequency region.

Table 2.21: Stream frequency (S_F) (Frequency distribution of sample basins)

Sl. No.	Drainage basins	Stream frequency (S_F) categories										Total frequencies
		Below-2		2-5		5-10		10-20		Above 20		
		F	%	F	%	F	%	F	%	F	%	
1	2	3	4	5	6	7	8	9	10	11	12	13
1.	Samrani	36	53.73	27	40.30	04	05.97	–	–	–	–	67
2.	Paroi	12	26.67	28	64.44	05	08.89	–	–	–	–	45
3.	Piprawal	07	18.42	27	73.68	04	07.90	–	–	–	–	38
4.	Barwa	03	06.52	16	34.78	10	21.74	16	34.78	01	02.18	46
5.	Ghinhal	03	07.32	07	17.07	20	48.78	11	26.83	–	–	41
6.	Jhuri	–	–	10	28.57	23	65.71	02	05.72	–	–	35
7.	Bagdara	07	13.21	26	49.06	20	37.73	–	–	–	–	53
8.	Kalibarah	02	04.88	13	31.71	15	36.59	11	26.82	–	–	41
9.	Maro	01	2.78	13	36.11	22	61.11	–	–	–	–	36
10.	Sukra	01	03.45	12	41.38	15	51.72	01	03.45	–	–	29
11.	Kulagara	–	–	06	33.33	11	61.11	01	05.56	–	–	18
12.	Pindrah	–	–	12	54.55	10	45.45	–	–	–	–	22
13.	Barar	–	–	12	38.71	18	58.06	01	03.23	–	–	31

(Contd...)

1	2	3	4	5	6	7	8	9	10	11	12	13
14.	Kewai	04	05.33	28	37.33	38	50.67	05	06.67	–	–	75
15.	Ganta	–	–	02	11.11	13	72.22	03	16.67	–	–	18
16.	Jhurai	07	29.17	12	50.00	05	20.83	–	–	–	–	24
17.	Gauhia	07	21.21	24	72.73	02	06.06	–	–	–	–	33
18.	Kachchua	25	67.57	12	32.43	–	–	–	–	–	–	37
19.	Dhora	20	50.00	16	40.00	04	10.00	–	–	–	–	40
20.	Dingchi	11	26.19	20	47.62	11	26.19	–	–	–	–	42
Total frequency and average frequency% =		146	18.94	323	41.89	250	32.43	51	06.61	01	00.13	771

Poor stream frequency (F_P) = 60.83%, Moderate stream frequency (F_M) = 32.43%

Moderately high stream frequency (F_{MH}) = 6.61%, High stream frequency (F_H) = 0.13%

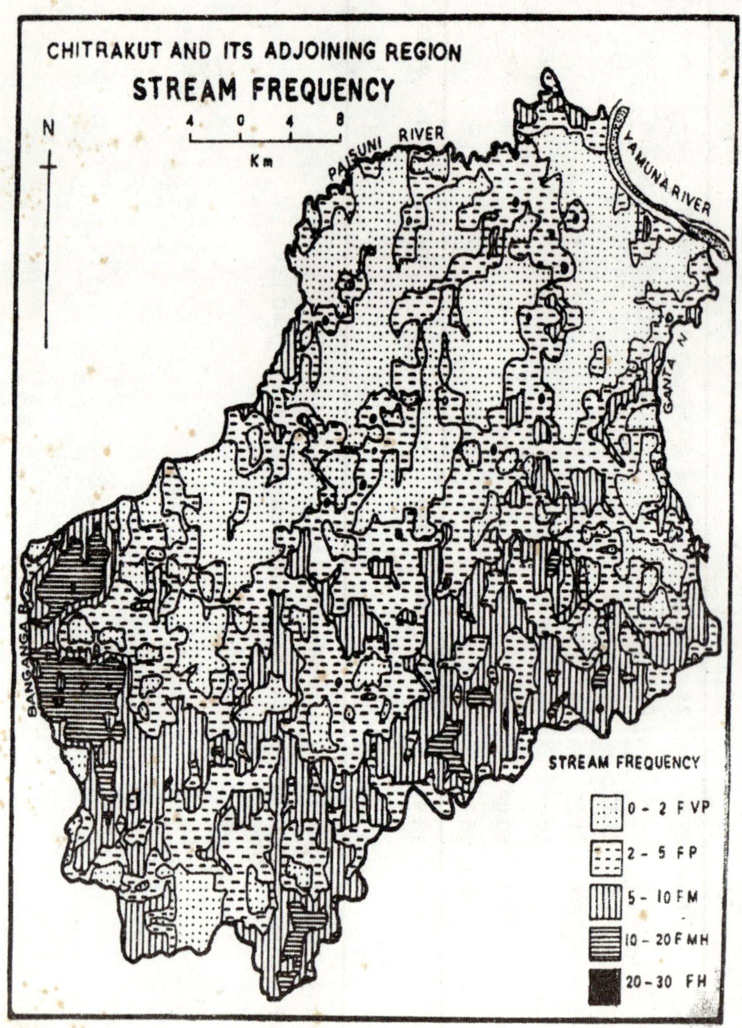
CHITRAKUT AND ITS ADJOINING REGION
STREAM FREQUENCY
N
4 0 4 8
Km
PAISUNI RIVER
YAMUNA RIVER
GANTA N
BANGANGA R.
STREAM FREQUENCY
0 - 2 F VP
2 - 5 F P
5 - 10 F M
10 - 20 F MH
20 - 30 F H

Fig. 2.19

In general, stream frequency provides more or less similar findings as depicted by drainage density and drainage texture. These three variables have largest percentage of areal coverage in their lower limit and the frequency percentage decrease as well as the range of limit increases. Thus the higher limit indicates the lesser frequency percentage. The isopleth maps of these three variables appear almost identical with minor variations. It is also remarkable that Dt and S_F are much closer to each other than that of Dd with Dt and S_F. The area in the north of 25° 10′ N latitude is found as a significant part of lower categories while the western fringe of southern plateau is treated as the region of high frequency categories for these three variables.

On an average, it may be concluded that drainage texture and stream frequency are more significant parameters of dissection of terrain than drainage density. The results of these three variables highlight that these are closely related to each others.

AVERAGE SLOPE

The study of slope form and its evolution has become the most important aspect of geomorphology. It is an angular inclinations of terrain between hill tops and valley bottoms and as an important component of the landscape which provides not only the variety of topographical features but also makes available the evidences needed for the interpretation of the complex form of landscape (Kumar, A. 1979). It is directly affected by several environmental factors and also influences few of them.

The present study is based on the computation of average slope from the topographical maps on the scale 1:50,000 with 20 m contour interval through Wentworth, C.K. (1930) method. The entire region is divided in grids of one km x one km and average angles have been computed according to the following formula:

$$\text{Slope angle} = \text{Tan } \theta = \frac{(N) \times CI}{0.6366K}$$

where N = number of intersections of contours per km.,

CI = vertical contour internal in meters, and

K = 1000 for metric units and 5280 for feet and miles.

On the basis of the above formula the slope angles in each grid have been obtained and classified in five categories. Table 2.22 clearly illustrates that the level slope category carries 46.48 per cent of the total frequencies of the region. If level and gentle slope categories are considered together, they exhibit 69.62 per cent of the total frequencies. Moderate, moderately steep and steep slope categories indicate 21.68 per cent, 7.51 per cent and 1.19 per cent of the total frequencies respectively. The infra-basinal study also produces similar findings. Thus the above explanation reveals the fact that the angle frequency of gentle slope is considerably more than steep slope because steep slopes are more rapidly consumed and altered by denudational processes than gentle slopes and thus the latter survives for a longer period of time than the former.

As regards the distribution of average slope frequencies in different slope categories of 20 sample basins (Table 2.23) it is obvious that maximum angle frequency (32.76%) in level slope category whereas gentle, moderate, moderately steep and steep slope categories carry 22 per cent, 31.93 per cent, 11.50 per cent and 1.81 per cent respectively. The basinwise study also illustrates the above fact. Isopleth map (Fig. 2.20) portraying spatial distribution of average slope reveals the fact that there are well marked variation in different categories of average slope angles. The major part of Yamuna plain located in the north of 25° 12′ N latitude is mostly characterized by level and partly by gentle slope categories. This is due to the continued operation of denudational processes wherein major streams like Paisuni, Ohan and Ganta have progressively affected regression of southern scarps by back wearing of the watershed. The second region of level slope is highly correlated with the great ravine tract of the region usually formed between the plain and plateau country. The water divide line specially flat top of the plateau in the north of Manikpur, the area surrounding the upper reaches of Paisuni in the extreme south are shaded with gentle slope category. Moderate slope is scarcely measured in the north plain along the valley sides or along the lower sections of isolated granitic hillocks or residual hills whereas it generally covers the entire lower section of the scarp-zones located in the south. Moderately steep slopes are observed extensively in several scattered patches just below the free face sections of the scarps while the steep slopes are confined to the free face of the scarp (Prasad, G. 1987).

Table 2.22: Frequency distribution of slope categories (regional and intrabasinal)

Sl. No.	Areal unit	Slope categories										Total	
		Below-2 (S_L) Level slope		2-5 (S_G) Gentle slope		5-10 (S_M) Moderate slope steep		10-15 (S_{MS}) Moderately slope		Above 15 (S_S) Steep slope			
		F	%	F	%	F	%	F	%	F	%	F	%
1.	Entire region	860	46.48	428	23.14	401	21.68	139	07.51	22	01.19	1850	100.00
2.	Intra-basinal	283	32.76	168	22.00	220	31.93	85	11.50	15	01.81	771	100.00

Table 2.23: Average slope categories (frequency distribution of slope for 20 sample basins)

Sl. No.	*Drainage basins*	*Slope categories*										*Total frequencies*
		Below-2 Level slope		*2-5 Gentle slope*		*5-10 Moderate slope*		*10-15 M. steep slope*		*Above 15 steep slope*		
		F	%	F	%	F	%	F	%	F	%	
1	2	3	4	5	6	7	8	9	10	11	12	13
1.	Samrani	63	94.03	04	05.97	–	–	–	–	–	–	67
2.	Paroi	25	55.56	06	13.33	11	24.24	03	06.67	–	–	45
3.	Piprawal	22	57.89	08	21.05	04	10.53	03	07.89	01	02.64	38
4.	Barwa	12	26.09	09	19.57	09	19.57	12	26.09	04	04.68	46
5.	Ghinhal	05	12.20	05	12.20	05	12.20	19	46.34	07	17.06	41
6.	Jhuri	–	–	01	02.86	24	68.57	10	28.57	–	–	35
7.	Bagdara	05	09.43	11	20.76	33	62.26	04	07.55	–	–	53
8.	Kalibarah	09	21.95	18	43.90	10	24.39	04	09.76	–	–	41
9.	Maro	05	13.89	10	27.78	17	47.22	04	11.11	–	–	36
10.	Sukra	02	06.90	09	31.03	16	55.17	02	06.90	–	–	29
11.	Kulagara	01	05.56	05	27.78	10	55.56	02	11.11	–	–	18
12.	Pindrah	01	04.55	08	36.36	13	59.09	–	–	–	–	22

(Contd...)

1	2	3	4	5	6	7	8	9	10	11	12	13
13.	Barar	05	16.13	16	51.61	10	32.26	–	–	–	–	31
14.	Kewai	22	29.33	29	38.67	24	32.00	–	–	–	–	75
15.	Ganta	–	–	01	05.56	13	72.22	04	22.22	–	–	18
16.	Jhurai	12	50.00	05	20.83	06	25.00	01	04.17	–	–	24
17.	Gauhia	23	69.70	06	18.18	03	09.09	–	–	01	03.03	33
18.	Kachchua	32	86.49	05	13.51	–	–	–	–	–	–	37
19.	Dhora	23	57.50	03	07.50	06	15.00	08	20.00	–	–	40
20.	Dingchi	16	38.09	09	21.43	06	14.29	09	21.43	02	04.76	42
Total frequency & Average F %		283	32.76	169	22.00	220	31.93	85	11.50	15	01.81	771

Gentle slope (S_G) = 54.76% Moderate slope (S_M) = 31.93% Moderately steep slope (S_{MS}) = 11.50% Steep slope (Ss) = 01.81%

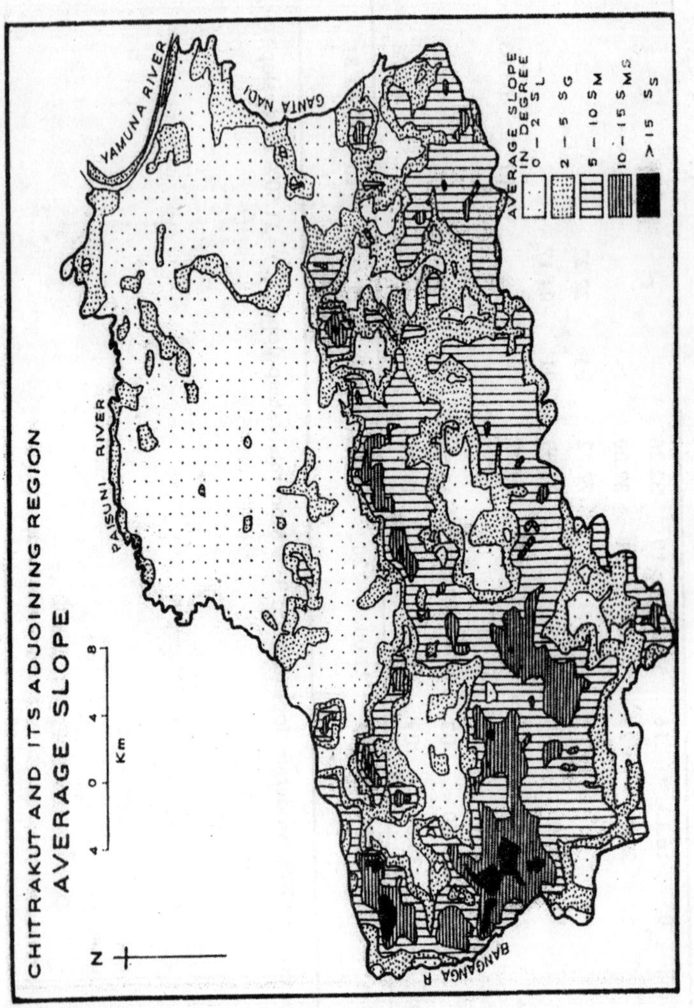
CHITRAKUT AND ITS ADJOINING REGION
AVERAGE SLOPE
4 0 4 8
Km
N
PAISUNI RIVER
YAMUNA RIVER
GANTA NADI
BANGANGA R
AVERAGE SLOPE IN DEGREE
0 — 2 S L
2 — 5 S G
5 — 10 S M
10 — 15 S MS
> 15 S s

Fig. 2.20

CLINOGRAPHIC ANALYSIS

Clinographic analysis has been proved one of the most important morphometric tool of paramount significance for the study of relationship between the height and slope angles between two successive contours. It also illuminates general slope profiles and designates the determination of breaks in slope profiles which help in constructing the morphological history of the region and in establishing the missing links of morphological events (Singh, U.P. 1984).

The clinographic analysis has been introduced by several geomorphologists but the following method of Strehler, A.N. (1952) has been applied in the present study:

Average slope angle between

two successive contours $= \text{Tan } \theta = \dfrac{CI}{AW}$

where CI = Contour interval; and

AW = average width between two successive contours.

When $AW = \dfrac{A}{(L_1 + L_2)/2}$

where A = area between two successive contours; and

L_1 & L_2 = lengths of the successive contours.

On the basis of the derived data, the clinographic and slope height curves (Fig. 2.21) have been drawn for the entire region and 18 sample basins. The clinographic curves of entire study region show a free face concave profile with limited rectilinear element just below the free face element. There are several breaks in slope at 140 m, 180 m and 380 m in the total length of profile wherein 140 m break heralds the termination of flat low land country in the north. It is interesting that the slope below the height of 260 m is convexo-concave in nature whereas the upper section of the profile is free face characterized with rectilinear element in their lower sections throughout the area. The trend of slope is rectilinear between the height of 260 m and 380 m and the vertical cliffing is most common above 380 m break. It may be pointed out that the concave element has its maximum length below the height of 180 m. The breaks in slope denote rapids and waterfalls. In the case of 18 sample basins, the clinographic curves reveal the fact that the basins like Piprawal, Jhurai, Gauhia, Dhora, Kewai, Barwa, Paroi and Dingchi generally indicate convexo-concave slope profiles with graded stage. Ghinhal basin exhibits total convexity whereas Bagdara basin produces a perfect rectilinear slope. Maro basin shows a rough rectilinear slope profile with several change of slope at the height of 220 m, 260 m, and 360 m respectively. The rectilinear slope profile of Jhuri river registers breaks at the height of 260 m and 300 m. Basins like Sukra, Kulagara, Pindrah, Kalibarah, Barar and Ganta have their convexo-rectilinear slope forms. Kalibarah gives a peculiar view indicating few breaks of slope in profile. Dhora, Gauhia, Jhuri, Piprawal and Paroi register break in slope at the height of 140 m which are the indicative of terminal points between upland and low land country. The profiles clearly indicate the nature of terrain development and breaks located on the profiles illustrates some classical turn in structural formulation and tectonic movement in the history of various stages of terrain development in the region.

DRAINAGE DISSECTION

Drainage dissection is a significant morphometric tool and as an indicator to estimate the nature and magnitude of dissection with respect to vertical exaggeration of terrain. It is expressed as the ratio between the maximum relative altitude and absolute height. It is dimensionless because it is always expressed in terms of ratio or percentage. The concept of relative relief and its importance becomes useless, if the intensity of absolute altitude may not differ in nature and thus the dissection becomes meaningless because equal relative relief is not always of equal importance since their absolute relief may not differ (Prasad, G. 1988).

Drainage dissection is calculated with the help of Deunir's (1957) formula given below:

$$D_I = \frac{R_R}{A_R}$$

where R_R = relative relief; and

A_R = absolute relief.

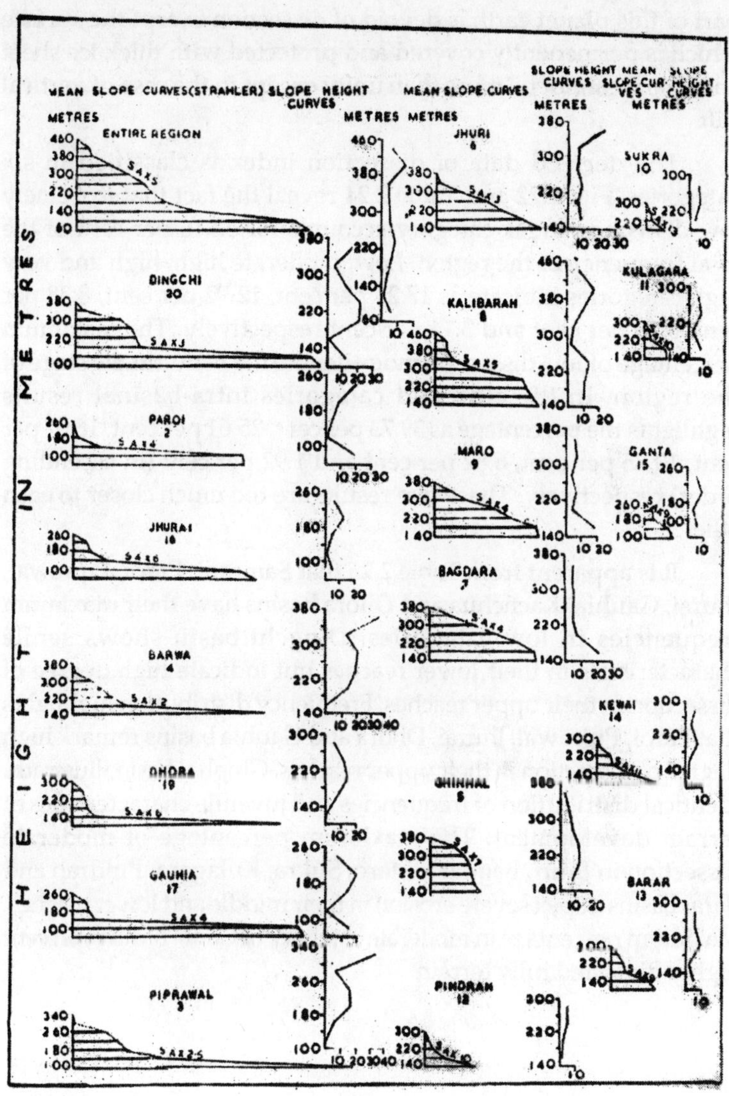
MEAN SLOPE CURVES(STRAHLER)
SLOPE HEIGHT CURVES
MEAN SLOPE CURVES
SLOPE HEIGHT CURVES
MEAN SLOPE CURVES
SLOPE HEIGHT CURVES
METRES
HEIGHT IN METRES
ENTIRE REGION
DINGCHI 20
PAROI 2
JHURAI 16
BARWA 4
DHORA 19
GAUNIA 17
PIPRAWAL 3
JHURI 6
KALIBARAH 8
MARO 9
BAGDARA 7
GHINHAL 5
PINDRAH 12
SUKRA 10
KULAGARA 11
GANTA 15
KEWAI 14
BARAR 13

Fig. 2.21

The values of dissection index are basically a ratio which range between O and O1 but it cannot be absolute zero because no part of this planet earth is devoid of dissection except the surface which is permanently covered and protected with thick ice sheet and also it cannot be more than unity except in the case of vertical cliff.

The derived data of dissection index is classified in six categories. Fig. 2.22 and Table 2.24 reveal the fact that extremely low dissection index category accounts for 52.16 per cent of the total frequency of the region. Low, moderate high, high and very high categories indicate as 17.24 per cent, 12.92 per cent, 8.38 per cent, 6.16 per cent and 3.14 per cent respectively. The maximum percentage of low dissection index denote the penultimate stage of the region. In the aforesaid categories intra-basinal results highlights the percentage as 39.73 per cent, 25.61 per cent, 16.07 per cent, 10.85 per cent, 6.82 per cent and 1.92 per cent (in ascending order) respectively. These two results are too much closer to each other.

It is apparent from Table 2.25 that Samrani, Paroi, Piprawal, Jhurai, Gauhia, Kachchua and Dhora basins have their maximum frequencies in low categories. Dingchi basin shows senile characteristics in their lower reaches but indicate high degree of dissection in their upper reaches. Frequency distribution highlights that Paroi, Piprawal, Jhurai, Dhora and Gauhia basins remark high degree of dissection in their upper reaches. Ghinhal basin illustrates identical distribution of frequencies and juvenile characteristics of terrain development. The maximum percentage of moderate dissection in Jhuri, Bagdara, Maro, Sukra, Kulagara, Pindrah and Barar basins depict severe erosion in their middle and lower reaches. Maximum percentage in moderate category of Ganta basin represent highly dissected hilly terrain.

Table 2.24: Frequency distribution of dissection index (DI) (regional and intrabasinal)

Sl. No.	*Areal units*	*Dissection index categories*											
		Below 0.1 (D_{EL})		*0.1-0.2* (D_L)		*0.2-0.3* (D_M)		*0.3-0.4* (D_{MH})		*0.4-0.5* (D_H)		*Above 0.5* (D_{VH})	
		F	*%*	*F*	*%*	*F*	*%*	*F*	*%*	*F*	*%*	*F*	*%*
1.	Entire region	965	52.16	319	17.24	239	12.92	155	08.38	114	06.16	58	03.14
2.	Intra-basinal	332	39.73	176	25.61	119	16.07	73	10.85	57	06.82	14	01.92

D_{EL} = Extremely low dissection index;
D_L = Low dissection index;
D_M = Moderate dissection index;
D_{MH} = Moderately high dissection index;
D_H = High dissection index;
D_{VH} = Very high dissection index

Table 2.25: Frequency distribution of dissection index (DI) (sample drainage basins)

Sl. No.	*Drainage basins*	*Dissection index categories*												*Total frequencies*
		Below 0.1		*0.1-0.2*		*0.2-0.3*		*0.3-0.4*		*0.4-0.5*		*Above 0.5*		
		F	*%*	*F*	*%*	*F*	*%*	*F*	*%*	*F*	*%*	*F*	*%*	
1	*2*	*3*	*4*	*5*	*6*	*7*	*8*	*9*	*10*	*11*	*12*	*13*	*14*	*15*
1.	Samrani	65	97.01	02	02.99	–	–	–	–	–	–	–	–	67
2.	Paroi	26	57.78	05	11.11	09	20.00	05	11.11	–	–	–	–	45
3.	Piprawal	27	71.05	01	02.63	–	–	01	02.63	08	21.06	01	02.63	38
4.	Barwa	10	21.74	08	17.39	04	08.70	03	06.52	17	36.95	04	08.70	46
5.	Ghinhal	11	26.83	07	17.07	08	19.51	10	24.39	04	09.76	01	02.44	41
6.	Jhuri	–	–	13	34.29	13	37.14	07	20.00	02	05.71	–	–	35
7.	Bagdara	10	18.87	28	52.83	11	20.75	03	05.67	01	01.88	–	–	53
8.	Kalibarah	21	51.22	09	21.95	08	19.51	–	–	03	07.32	–	–	41
9.	Maro	09	25.00	17	47.22	08	22.22	02	05.56	–	–	–	–	36
10.	Sukra	09	31.04	10	34.48	10	34.48	–	–	–	–	–	–	29
11.	Kulagara	02	11.11	11	61.11	05	27.78	–	–	–	–	–	–	18
12.	Pindrah	04	18.18	12	54.55	05	22.73	–	–	01	04.54	–	–	22
13.	Barar	06	19.35	17	54.84	07	22.58	01	03.23	–	–	–	–	31

(Contd...)

1	2	3	4	5	6	7	8	9	10	11	12	13	14	15
14.	Kewai	18	24.00	23	30.67	20	26.67	12	16.00	02	02.66	–	–	75
15.	Ganta	–	–	01	05.56	04	22.22	12	66.67	–	–	01	05.55	18
16.	Jhurai	14	58.33	03	12.50	–	–	07	29.17	–	–	–	–	24
17.	Gauhia	20	60.61	07	21.21	–	–	03	09.09	–	–	03	09.09	33
18.	Kachchua	36	97.30	01	02.70	–	–	–	–	–	–	–	–	37
19.	Dhora	24	60.00	–	–	04	10.00	02	05.00	06	15.00	04	10.00	40
20.	Dinghchi	19	45.24	03	07.14	03	07.14	05	11.90	12	28.58	–	–	42
Total F and Average F %		331	42.93	178	23.09	119	15.43	73	09.47	56	07.26	14	01.82	771

Low drainage dissection (D_L) = 66.02%; Moderate drainage dissection (D_M) = 15.43%

Moderately high drainage dissection (D_{MH}) = 9.47%; High drainage dissection (D_H) = 9.08%

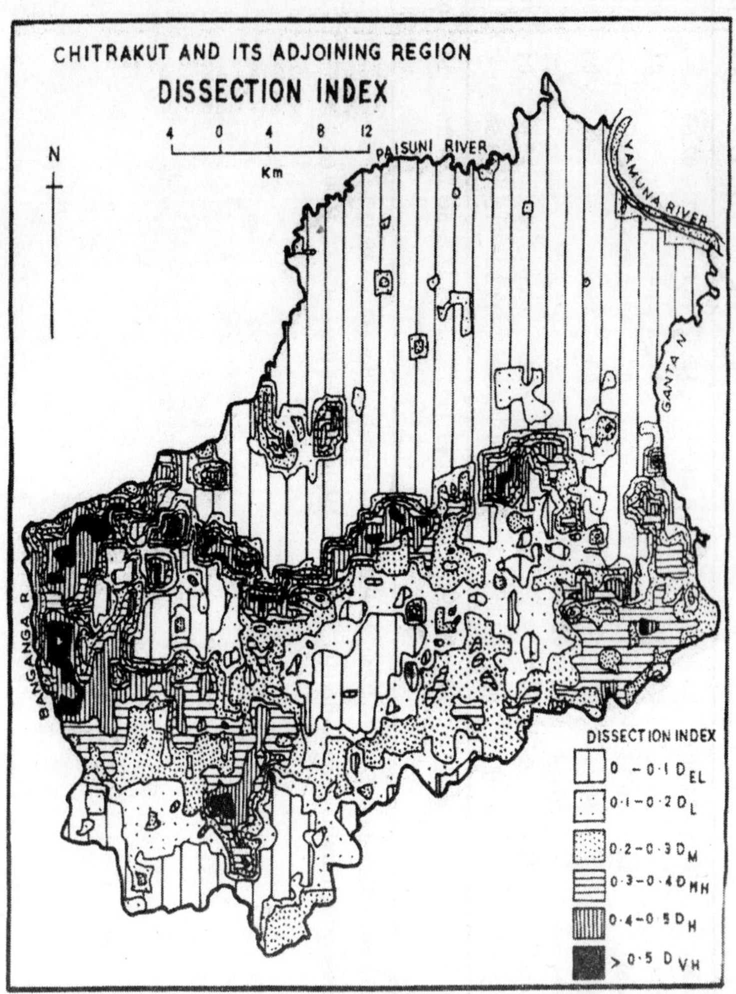
CHITRAKUT AND ITS ADJOINING REGION
DISSECTION INDEX
4 0 4 8 12
Km
N
PAISUNI RIVER
YAMUNA RIVER
GANTA N
BANGANGA R
DISSECTION INDEX
0 – 0·1 D EL
0·1 – 0·2 D L
0·2 – 0·3 D M
0·3 – 0·4 D MH
0·4 – 0·5 D H
> 0·5 D VH

Fig. 2.22

Fig. 2.22 clearly indicates that the very high dissection is scarcely marked in patches along western hill ranges below the height of 220 m. It is apparent that the degree of dissection has been decreased from scarps to northern flat land. The hill ranges generally show all the categories of dissection increasing from foothill to free face element. The scattered hill ranges near Chitrakut Dham also determine high degree of dissection. From Brahmkund nadi to Khoh Pahari (Karwi) the length of the scarps is closely surrounded by low to very high dissection. The area north of 25° 10′ N latitude is less dissected.

3

Intensity Measurement of Morphological Processes

PROCESS RESPONSE INTRODUCTION

Dutch ecologist Van Leewen (1966) introduced the terms 'Pattern' and 'Process' respectively which are of special importance with respect to the introduction of cybernetical principles in ecology for expressing the different kinds of relationships. Pattern indicates the spatial variation in a community at a given moment whereas process indicates the variation during a period of time. Both are evident in the development of a community; physical and cultural. The growing eco-system complexity develop the climax form of pattern and processes either internal or external. The above two determine the space-time relationship and variability of so many factors of eco-system, shaping, interacting and changing the entire framework. Processes disclose the major concept of geological impact, climatic complexity and other endo-exo-dynamics in the smiling face of the landscapes. Thus, the study of processes; its impact and intensity through a passage of time and in terms of spatial relationships are too much necessary for geomorphologists. The geomorphologists can judge the landforms by processes and processes can be viewed well by landforms. The space and time exist very important for both landform and process. The intensity makes clear the view of process response (Jayaswal, S.N.P. and Prasad, G., 1989).

PROCESS TYPES

The geomorphic processes are all those physical and chemical changes which effect a modification of the earth's surficial form. A geomorphic agent or agency in any natural medium which is capable of securing and transporting earth material (Thornbury, W.D., 1954). The running water (including both concentrated and unconcentrated runoff), ground water, glaciers, periglacier, wind and movements within water bodies including waves, currents, tides and Tsunami are the great geomorphic agencies designated as mobile agents. They remove, transport and deposit the materials at different parts of the earth surface.

Penck advocated that the agencies and the processes performed outside the earth crust have been termed as exogenetic. The other geomorphic processes which originate within the earth's crust are nominated as endogenetic.

The following is an outline of the processes which shape the earth's surface—

Geomorphic Processes

Epigene or Exogenetic processes

Gradation

Degradation

Weathering

Mass wasting or gravitative transfer

Erosion (including transporting) by:

Running water

Ground water

Waves, currents, tides and Tsunami

Wind

Glaciers

Aggradation by:

Running water

Ground water

Waves, currents, tides and Tsunami

Wind

Glaciers

Work of organisms including man hypogene or Endogenetic processes

Diastrophism

Vulcanism

Extra-terrestrial processes

Infall of Meteorites

The composite form of weathering and erosion is termed as denudation. In the case of erosional agents the river actions are designated as fluvial processes.

BASIS OF MEASUREMENT OF MORPHOLOGICAL PROCESSES

Process defines the dynamic actions or events in geomorphological systems which involve the applications of forces over gradients. These actions are caused by the factors like the wind and falling rain, waves and tides, river and soil water solutions. The changes occurred in the original form of the earth's surface indicate the operation of such processes.

The fundamental task of geomorphology is to understand the relationships between form and processes. They may be expressed in a cause-and-effect chain or for example, by the rates of change of operation of one with respect to rates of change of the other (Thornes, J.B., 1980).

Geomorphologists consider processes at various levels of spatial resolution. At the finest level of field observation, one may concern with the filling of pore spaces in the soil by moving water. The thermal convection below the earth crust is an example of a process at a very coarse scale. In the recent period, the geomorphologists have chosen the geomorphic unit for the measurement of the processes mainly of the order of a few hundred square kilometres or less. Processes also occur at varying rates. A force may be applied rapidly or slowly; its magnitude may be large or small. If we observe a particular form or assemblage of forms over a short period, there may be little change either because the forces are small or because they have not operated for a sufficient length of time. If the nature or rate of operation of the processes change there may be a corresponding change in the landforms (Thornes, J.B., 1980).

The investigations of process; its nature, magnitude and intensity are deeply concerned with in three framework:

1. The investigation of an evolution of forms in terms of an evolution of processes. The main intellectual thrust of this activity has been in determining the sequence of forms and deposits. The assumption is that the processes are understood.
2. The measurement based on the spatial and temporal variations of the processes and particularly on the problems of the magnitude and frequency of events.
3. The measurement initiated to focus mainly the relationships between forms and processes.

It has been generally recognised that the problems existing during the course of investigation of processes in the field, laboratory and office are essentially different. Certainly, all these modes of working are necessary if advances are to be made within the frameworks outlined above.

The morphological processes can be measured on the basis of the following techniques:

Field Work

Field work is crucial to studies of process because it confronts the investigator with associations and events of geomorphological significance which are outside his range of experience and hence prompt his curiosity. It was this type of confrontation together with a steadiness of thought and accuracy of observation that made Gilbert one of the greatest pioneers in process geomorphology. The same confrontation led Leopold to his remarkable investigations of fluvial processes of forms (Leopold, L.B., 1978). In a true sense, the field work can be an informal, sometimes almost incidental activity. But it is necessary and very important to the geomorphologists because it provides the field data for (i) the critical analysis of empirical relationships; (ii) the parameterization of theoretical models; and (iii) the testing and verification of existing theoretical concepts. These are or should be much more formal activities based on carefully developed field programmes, in which objectives and procedures are precisely defined.

Through the chronicles of field measurement, most commonly it is observed that processes have been inferred from the close-up view of an existing form or change of form in the field. Agassiz and Gilbert, the early workers of process studies, laid the foundations of modern process studies by direct field measurement combined with seemingly effortless induction of the large scale processes operating over wide areas of the earth surface. Upto the first half of twentieth century inferences were made from field evidences about the nature and rate of operation of processes. The inferences drawn by Wayland, E.J. (1947) as to the processes responsible for tropical weathering from the extensive plains of low latitudes and Kings, L.C. (1962) about the occurrence and magnitude of sheet wash of peneplains are notable. These inferences became the part and parcel of geomorphological thinking because they led to the development of hypotheses that can be tested in a more rigorous framework. The comparison of river-channel forms before and after floods (e.g., Thornes, J.B., 1976, Anderson, M.G. and Calver, A., 1977) and upstream and downstream from reservoirs (Gregon, K.J. and Park, C., 1974) are the recent examples in the case of processes measurement. There exist a major problem to recognise the relationship between form and process specially in the above procedure. The varying processes give rise to multidimensional forms. It is also worth considering that the same process may result widely differing forms through the variability in space and time. During the course of field observation, it may be minutely observed that the form-process relationship is heavily conditioned and controlled by structure, lithology and in most places by human activity. Thus the processes can be measured by careful and well designed experiments. The variability of the effects of different processes that occurs at one place in time is the major difficulty when inferring process from form. The processes like weathering, soil creep and erosion by running water occur together on most of the hillsides. The climatic variations indicate sudden swift on one landscape. Such changes, most probably, may be recognised in different ways of geomorphologist's understanding.

A field offers a primer, a challenge to our explanatory powers, a realistic tableau of what we have to explain rather than the explanation itself (Thornes, J.B., 1980). The recent trend of geomorphology has compelled the geomorphologists to validate the need of the process models which have been developed to identify them. The validation involves two operations: (i) the appreciation of the likely domain over which process inputs operate in terms of magnitude and frequency; and (ii) the determination of constants (parameters) in physical relationships. The verification involves checking the expected results of process models against the field data. One can mention the most successful examples of field work completed on the mechanics of glacier movements (e.g., Meier, M.F., 1960), the evaluation of hydrological processes in small catchments such as the Waggon Wheel Gap experiments (Bates, C.G. and Henry, A.J., 1928) and studies of process rates in the coastal zone (Mcclean, R., 1967).

The sampling problems, the cost of providing and maintaining expensive instrumentation and the severe technical constraints linked with the constant need to interfere with the processes are the major problems existing under observation. Apart from the above difficulties, the field work can provide the best possible results for the nature and intensity of the processes operating at any place.

Laboratory Investigations

The field observation provides relevant data on the spatial and temporal ranges of inputs and parameters for models. It discloses the inter-relationships among complex processes and gives a central idea of understanding of process and form relationships.

In the laboratory, the emphasis is on the control of events. But the reality of interaction of land forming processes cannot be truly examined. The laboratory provides the solution to scale down the space and speed-up time. It gives a chance to prepare a theoretical model of a terrain or the operating processes.

In the case of hydraulic processes, laboratory studies offer an excellent result. The work of Griggs, D.T. (1936b) on arid zone weathering, Lewis and Miller (1955) on model glaciers, and by Van Burkalow, A. (1945) on mass movement is the fundamental work based on laboratory analysis. One of the most exciting and largest scale experiments in recent years has been Schumm's attempt to produce an entire drainage basin system under cover (Schumm, S.A., 1973).

The laboratory analysis relates to the mass length and time properties of the model and to the scaled-down versions of the real-world processes. Sprinkler device of erosional phenomena (Mosely, M.P., 1973) and the study of effects of freeze-thaw on rock particles (Potts, A.S., 1970) are the examples of above models:

Office Work

Investigations of processes by 'arm-chair' geomorphologists take the following three forms:

1. The analysis of field data in order to investigate empirical relationships.
2. The development of theories based on fundamental physical principles.
3. The development of laboratory-type experiments on digital or analogue computers.

All three levels have made highly significant contributions to the understanding of geomorphological processes (Thornes, J.B., 1980).

An empirical and theoretical work has been done through office work. The major orientations have been drawn on the relationships between factors or agents and the response variables through multiple regression and correlation analysis in empirical attempts. This attempt has become more sophisticated by incorporating temporal and spatial lags in estimated relationships. This attempt was initiated by Strahler, A.N. (1954) and Melton, M.A. (1957, 1958) wherein multiple regression techniques have been widely used in fluvial process studies. Walling, D.E. (1971) introduced a four variable regression model for fluvial measurement. The problem of availability of data, correlation and further assessment about processes are the important question of this attempt. In theoretical work of office study, the nature and characteristics of the processes and its impact on the behaviour of landforms change are the subject of basic interest. In this field Thornes, J.B. (1978) reviewed all the previous work and the later developments are mostly made in the field of hydraulics and river sedimentation. Young's, A. (1963) study in the case of slope evolution is an excellent example in sense of theoretical interpretation. R.E. Horton's (1938) study in the case of discharge, surface flow, uniform rate of rainfall and water supply rate are also notable in this field.

A major element of theoretical work and one that causes some difficulty is the transportation of a verbal statement into a symbolic statement usually mathematical. This is because the mathematical formulation eases the process of education and minimizes the likelihood of ambiguities. Now the processes under investigation may be easily simulated on a digital computer. This operation has the added advantage that elements of uncertainty in the inputs or parameters can be fairly easily accommodated. The computer analysis has provided more satisfactory services in this field. Ahnert, F. (1976) has proposed a complex computer programmes that simulates the effect of the processes on a three dimensional hypothetical land surface. Many other programmes of processes measurement have been introduced earlier. The present research plan is based upon the field measurement, laboratory analysis and office work. Seasonal change of the nature of processes have been drawn through field study taking spot photographs. Office work includes the study and interpretation of topographical sheets and some other mathematical derivations. The testing of the rocks is included by laboratory analysis. Thus, all the facts and findings are depicted through three dimensional observations.

INTENSITY CHARACTERISTICS EVALUATION

The study are experiences the subtropical climatic conditions under which different landscapes have been formed due to dynamic cycle of fluvial processes associated with mechanical and chemical weathering. Due to lack of advanced instrumental facilities and some other mathematical techniques to measure the nature and magnitude of morphological processes, the intensity of the processes operating on the terrain of the region, has been determined with the help of suggested models of Peltier, L.C. (1950) and Wilson, L. (1973) depending upon the mean annual temperature (26.02°C) and mean annual rainfall (95.65 cm) of the region.

According to Peltier's model seven graphs showing the relationships between various geomorphic processes and climatic parameters of mean annual precipitation and temperature have been drawn and the location of the study region is marked with a dot (Fig. 3.1). Thus, the various types of weathering (Fig. 3.1A), the intensity zone chemical weathering, mechanical weathering, mass movement, fluvial erosion, aeolian action and the recognition of morphogenetic region incorporate the following results:

1. Chemical weathering—Moderate intensity;
2. Mechanical weathering—Insignificant;
3. Mass movement—Moderate intensity;
4. Fluvial erosion—Maximum intensity;
5. Aeolian action—Moderate intensity; and
6. Morphogenetic region—Moderate in nature.

On the basis of the above findings, it may be concluded that the fluvial processes are most dominant and play vital role in supporting various processes and forming as assemblage of landforms in the region. Mass movement which is defined as mass translocation of rock wastes mainly by flow, slide or subsidence/heave is of moderate nature. Aeolian actions are also seen in moderate intensity as explained in Fig. 3.1F. But this evaluation is contrary to the fact. Similarly the abnormal result is also found in the case of mechanical weathering because the insignificant impact of mechanical weathering is not possible in the region where fluvial erosion is much dominating.

On an average, the region falls under moderate morphogenetic characteristics wherein average annual temperature ranges between 38°F and 85°F and rainfall varies between 35" and 65" which are the indicative of maximum effect of running water and moderate mass movement with less effective wind actions.

The abnormalities present in the Peltier's model have been rectified in the plan of Wilson (Fig. 3.2). It is remarkable that the Wilson's model differs from figures produced by other workers (e.g. Peltier, L.C. 1950), Leopold, L. 1964, et al.) mainly in the emphasis given to desiccation processes. Fig. 3.2 illustrates the five determinants of morphological processes e.g. mechanical weathering, chemical weathering, mass wasting, running water and wind erosion respectively. It is apparent that almost similar findings have also been reported through the Wilson's graph for

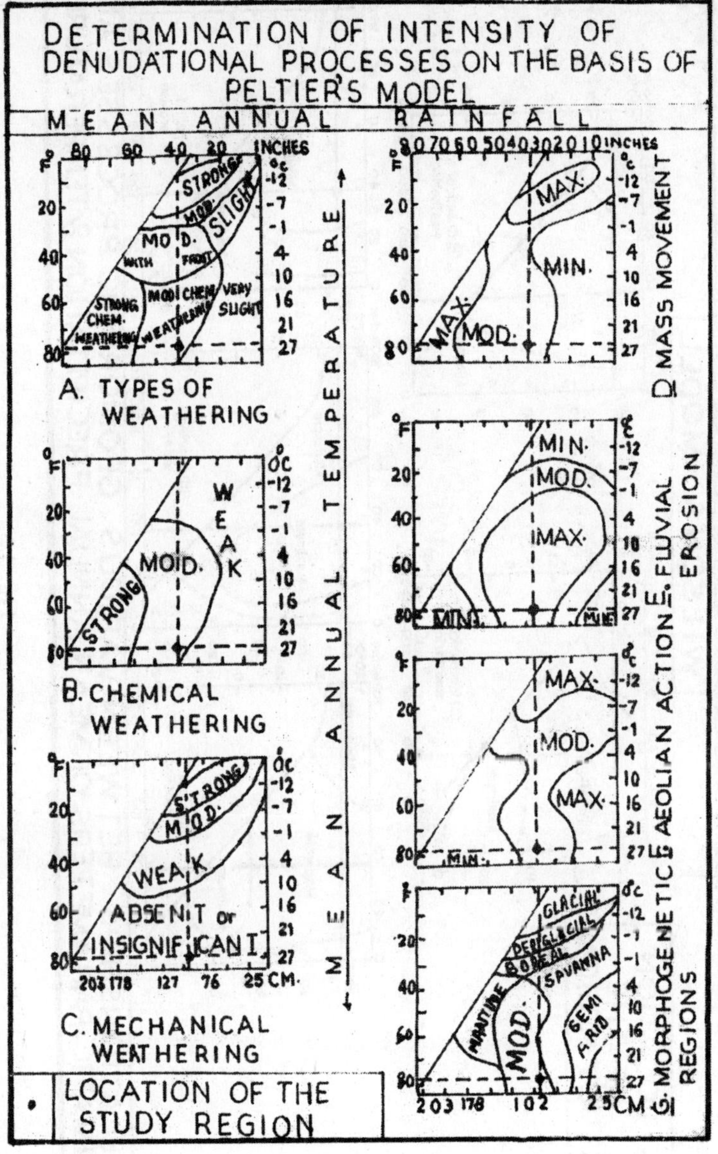
DETERMINATION OF INTENSITY OF DENUDATIONAL PROCESSES ON THE BASIS OF PELTIER'S MODEL
MEAN ANNUAL RAINFALL
MEAN ANNUAL TEMPERATURE
A. TYPES OF WEATHERING
B. CHEMICAL WEATHERING
C. MECHANICAL WEATHERING
D. MASS MOVEMENT
E. FLUVIAL EROSION
F. AEOLIAN ACTION
G. MORPHOGENETIC REGIONS
LOCATION OF THE STUDY REGION

Fig. 3.1

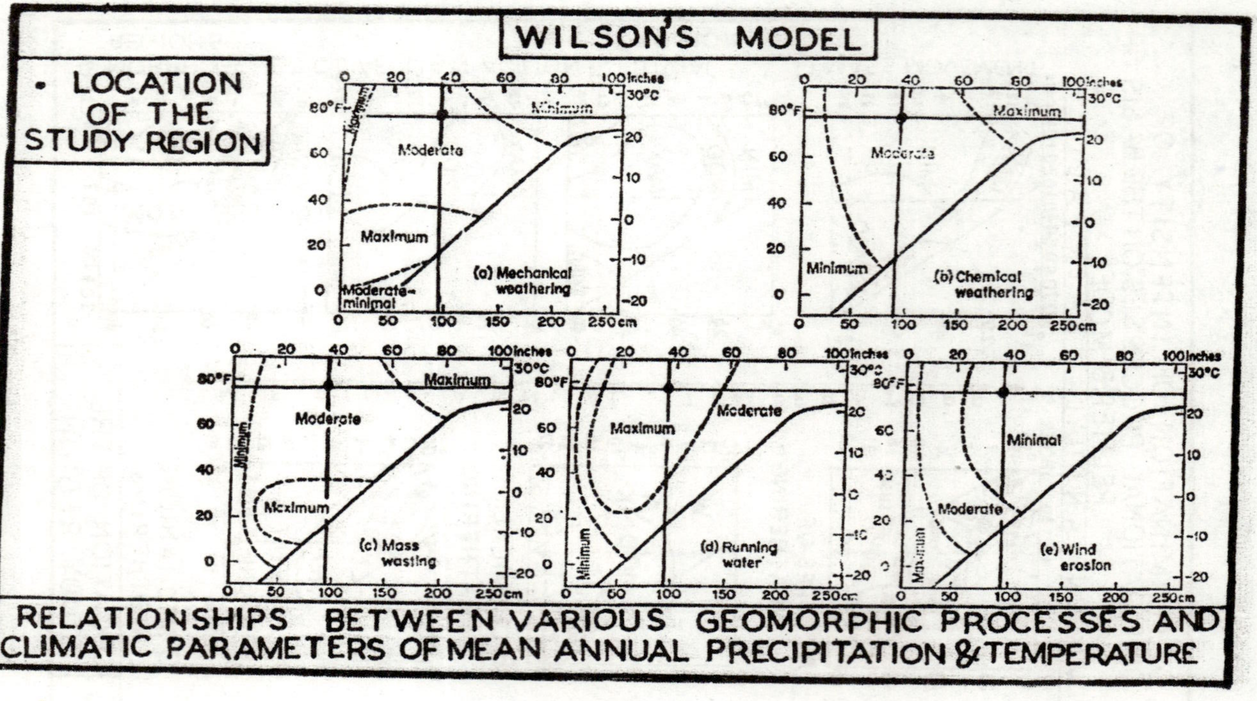
WILSON'S MODEL
• LOCATION OF THE STUDY REGION
(a) Mechanical weathering
Minimum
Moderate
Maximum
Moderate-minimal
(b) Chemical weathering
Maximum
Moderate
Minimum
(c) Mass wasting
Maximum
Moderate
Maximum
Minimum
(d) Running water
Maximum
Moderate
Minimum
(e) Wind erosion
Minimal
Moderate
Maximum
80°F
30°C
100 inches
250 cm
RELATIONSHIPS BETWEEN VARIOUS GEOMORPHIC PROCESSES AND CLIMATIC PARAMETERS OF MEAN ANNUAL PRECIPITATION & TEMPERATURE

Fig. 3.2

chemical weathering, mass wasting and dominance of running water. But the anomalies found in the evaluation of mechanical weathering and wind erosion have been totally removed because the mechanical weathering and the impact of wind erosion have been marked at moderate and minimal intensity respectively and provide satisfactory results about the intensity of morphological processes of the region.

Thus, the Wilson's model is more appropriate and applicable because it provides a clear picture of the relationships between various geomorphic processes and climatic parameters of the region.

CONCLUSION

Overall discussion reveals the fact that the intensity of the operating processes can be minutely observed through three-way analysis for example field measurement, laboratory treatment and office work. The measurement of processes intensity holds more satisfactory results adopting Wilson's model in comparison of Peltier's graph. For a better understanding of the processes involved, we should study also the climatic changes in the spatio-temopral context.

4

Weathering Processes and Mass Movement

INTRODUCTION

The term "weathering" which gained wide currency in literature does not reflect the substance and the totality of natural processes defined by this concept. There is no unanimity among researchers as to its essential meaning because of the unfortunate choice of the term. At any rate weathering should never be confused with the activity of wind as much as of other exogenous factors.

Weathering is an aggregate of processes involving physical destruction and chemical decomposition of minerals and rocks at the site of their occurrence caused by the variations in temperature, chemical action of water, gases, oxygen and carbon dioxide (contained in the atmosphere and dissolved in water) by the biochemical action of organisms in the course of their vital activity and by the products of their decomposition after they have died off (Siffer, V.V. 1977).

In view of Ollier, C. (1969), weathering is the breakdown and alternation of materials near the earth surface to products that are more in equilibrium with newly imposed physio-chemical conditions. The rocks found at or near the earth's surface are of completely different environments from those in which they originated because their morphogenesis have been subjected to high temperatures or high pressures in the absence of air and atmospheric water. This process of adjustment is called weathering and is the combined action of all processes which cause disintegration and decomposition of rocks to products that are more in equilibrium with their present environment (Clowes, A. and Comfort, P. 1982).

Perhaps the most widely accepted definition is of Reiche, P. (1950) who advocated that weathering is the response of materials which were in equilibrium within the lithosphere to conditions at or near its contact with the atmosphere, the hydrosphere and perhaps still more importantly the biosphere. Keller, W.D. (1957) is also of the opinion that Reiche's definition could be improved by deleting the words 'which were in equilibrium' for he says rocks are in equilibrium only momentarily, while the environment in which they formed persists.

Weathering has been described by Polynov, B.B. (1937) as 'the change of rocks from the massive to the clastic state'. It is pertinent to point out that weathering is not simply the reduction of minerals to fundamental particles. The formation of new minerals in a state of weathering profile is also considered as a part and parcel of weathering. Diagenesis which is most common term to weathering is the alteration of sediments by building up new minerals. Weathering is a phenomenon of the land and diagenesis is dominant in the sea.

The 'Chitrakut Upland' is largely covered with horizontally bedded sandstone and limestone. The rocks most probably have come to existence due to specific thermodynamic conditions in the interior of the earth, in the zones of magmatic activity and metamorphic processes. The physio-chemical conditions of the region have resulted the disintegration of rocks into fragments of different size and shape and into the separate constituent minerals. Majority of numerous chemical agents produce radial changes in the original forms of the minerals and rocks. Depending on the operative factors and the results of their operation, weathering processes are roughly subdivided into the two heads: Physical and Chemical. The two types are most intimately interwoven, act together and at the same time and only the intensity of their manifestation is different. It is because of dynamic change in climatic conditions, relief tectonics, time, the composition of the rocks and some other environmental factors of the region.

Organic weathering is distinguished as a third type of weathering. The physical or chemical processes played by the organisms on the rocks define this type of weathering. Plant and animal influences (Biogenesis) and human interferences (Techno-functions) cause the biotic weathering. The crux and clues of weathering processes as observed in the field study of the region are given systematically in the preceding paragraph.

PHYSICAL WEATHERING

Physical weathering is the breakdown of material by entirely mechanical methods brought about by a variety of causes. Some of the forces originates within the rock while others are applied externally. The applied stresses lead to strain and eventually to rupture (Ollier, C. 1969). According to Clowes, A. and Comfort, P. (1982), physical weathering involves the mechanical breakdown of rock into smaller particles by the exertion of stresses sufficient to strain and eventually split it.

Variety of factors introduce physical weathering, but the decisive role is played by those initiating mechanical movement of rock particles, which disturbs the mechanical bond between the rock constituents. These factors of physical weathering are related to a large extent by climate. The crustal growth within a rock, thermal expansion and contraction, pressure release and unloading and multidimensional biological interferences are the principal cause of physical destruction of earth crust.

The following processes relating to physical weathering are recognised during the extensive field study of the region.

Sheeting

Sheeting is the division of rock into 'sheets' or 'beds' by joint-like fractures generally found parallel to the ground surface. This process is also termed as the name of 'Topographic jointing'. Sheeting appears during the period of uplift and erosion when the expansion of rock masses suddenly develop the cracks. The sheeting is first recognised by Gilbert, G.K. (1904) in granite country and further surveyed by Bradley, W.C. (1963) in massive sand stone, Ollier, C. and Tuddenham, W.G. (1962) in massive arkose and conglomerate, Kiersch, G.A. and Asce, F. (1964) in limestone and by Currey, D.T. (1968) in bedded sandstone. Some other examples of sheeting are noticed by Chapman, C.A. and Rioux, R.L. (1958), Lewis, W.V. (1954), Oen (1965) and Gage, M. (1966) mostly in glaciated regions located in the different parts of the world.

In the case of the present study region, sheeting is noticed over standstone and limestone structure. It is due to overburden pressure released during the early period of tectonic movement. This type of examples can be easily traced near Sati Anusuiya Ashram where sheeting is formed by unloading. It is also observed at Gupt-Godavari hills. The surrounding area of Kamad Giri Parbat the sloping country of Nilgarh Pahar and the stony surface of Manikpur hills clearly indicate the example of sheeting. In all the cases, it is observed that the sheeting of sandstone and limestone surface is formed with pressure release cracks.

Unloading

There are minute differences between sheeting and unloading. Unloading causes sheeting. In the case of unloading, horizontal and sub-horizontal partings of rocks can be recognised generally towards the top of the exposures. It is notable that the unloading also results vertical cracks but it is found less dominating. These fractures suddenly appeared after release of confining pressure. Unloading is progressed by erosional agents.

Spalling

Spalling is the process which produces spalls of platy rock fragments generally lozenge-shaped or irregular. They break from the wall by a combination of tension cracks parallel to the unloaded surface and shear fractures due to compression acting parallel to the wall. Following to Ollier, C.D. (1969) spalling exists due to physical conditions favour cracking and by spatial considerations which prevent the formation of sheets. In spalling, the unloading produces stress differences which in turn cause uniaxial, biaxial and radial bending stresses. The field measurement shows that each of the bending stress easily can be found. Uniaxial and biaxial bending stresses are marked due to spalling process. The various parts of Gupta Godavari caves generally depict radial bending stresses due to spalling process. Here it is remarkable that sheeting, unloading and spalling processes jointly act at one place. These are interdependent to each other.

Block Separation or Disintegration

Apart from the above pressure release processes, weathering must be recognised due to physical properties of rocks. All the sedimentary rocks contain bedding planes separated horizontal or nearly horizontal layers of sediment as they were laid down. Thus, after the rocks have been uplifted and exposed in their present day position, bedding planes can function as openings or cracks which the weathering processes, whatever they may be use. Some of the rocks like limestone also have vertical cracks or joints from one bedding plane to the other. In the sedimentary rocks weathering enlarges the rock joints and so more of the rock surface exposes to the atmosphere suddenly breaking up the rock under block separation process. The above disintegration causes by the physical properties of the rock.

This type of block disintegration can be recognised over sandstone and limestone structure of the region. During the Gupta-Godavari cave survey, the separations are measured along the side wall. Paisuni river bed near Sati Anusuiya Ashram shows various blocks of limestone minutely separated to each other. The granite country and sandstone structure also illustrate block separation.

Spheroidal Weathering

Some well-jointed rocks, such as sandstone, limestone and granite decompose readily; the corners and edges are attacked more than the flat surfaces. Thus, the rock becomes rounded in all three dimensions. As each 'shell' is weathered away, a fresh 'shell' of rock is exposed and the rock decomposes, becoming more sphere-like as weathering progresses. This type of weathering process is called as 'spheroidal weathering'. In this condition the rocks show globular jointing. The surrounding area of 'Kamad Giri' hill indicates the 'Globular jointing' of granite rocks and the examples of spheroidal weathering. It can also be seen along the boulders controlled sloping surface of Pahari hills near Pahari development block. The big size rounded boulders along the Allahabad-Satna railway line about 3 km away from Manikpur railway station clearly define the case of spheroidal weathering.

Columnar Weathering

Columnar jointing in granite and sandstone rocks initiates Columnar weathering. Rocks contract and jointings are generally found in vertical manner. The weathering process attacks the corners and edges of the rocks most probably in the upper portions. The boulders in the eastern and western flanks of 'Kamad Giri' hills exhibit Columnar Weathering.

Hummock Weathering

Hummock jointings of sandstone and granite rocks form hummock weathering processes. The sandstone blocks appeared near Manikpur hills are the best examples caused by hummock weathering.

Weathering Processes Due to Crystal Growth

Volume changes due to crystal growth set up stresses within rocks that lead to breakdown and thus constitute a part of physical weathering. The changes of volume are examined due to freezing of water to form ice, and crystal growth from solution (salt weathering) or chemical alteration of pre-existing minerals. The process of frost shattering and frost wedging have not observed due to opposite climatic conditions but some cases of *rock swelling, granular disintegration* and exfoliation are seen in limestone formation of Gupta-Godavari caves. It is assessed that the crystallisation of salts within the pores at the surface of limestone rock may lead to granular disintegration. But in a true sense, it is not correct because the salt content in limestone area is not possible. In the case of river valleys and their surrounding flood plains, it is undoubtedly correct. The river side wet area generally shows swelling conditions and sometime chips like sheet has been found bodily away from the upper horizon of the soil.

Exploitation

Temperature changes cause expansion or shrinkage of rocks. The variation in temperature during the day and night causes non-uniform heating and cooling of the rocks which in turn greatly influences the volume of consolidated and fragmental materials to break up. In the case when heating agent is sunshine the said process is termed as insolation weathering. The heating and cooling generally depend on the nature of the rocks. The continued process of temperature change initiates unequal expansion and construction of the rock mass and further intensifies the integration of rocks or peel away the surface layers from the parent rock. This phenomenon is known as exploitation or destruction and desquamation of rocks.

Such type of weathering process is measured over Nilgarh Pahar near Bharatpur village where a rounded shaped boulder is marked with two set of cracks, one parallel to the boulder surface and the other vertical in nature. Although it is sun-cracked boulder but the major role has been played by dirt intrusted within the boulder. Geomorphologists also recognised this type of boulders as dirt cracked boulders.

Reiche, P. (1950) has elaborated the impact of boulder cleaving over dark and fine grained rocks due to thermal expansion of the outer heated layer of the boulder during sunnydays. This type of boulder cleaving caused by insolation weathering is noticed over Nilgarh Pahar near Bharatpur village and Satrajkifa Pahari near Pathnauri village, Sheorampur.

Slaking and Splitting

Alternate wetting and drying of rocks can be a very important factor in weathering, a process known as slaking. The two types of disintegration of rocks may be possible, a minor disintegration and a major disintegration of rocks. The minor is caused due to slaking and the major is initiated by splitting in two or more large pieces of about equal size. Cracking generally intended to concentrate along the bedding planes and cleavage planes when these are formed. This type of weathering processes can be observed in fine grained rocks. The limestone and sandstone surface near Sitakund indicates few examples of slaking and splitting. The flowing water of river Paisuni and sunnydays have provided alternate cooling and heating to the said rock surface.

Mechanical Collapse

Rock weathering initiates the form of mechanical collapse following undercutting of various sorts. Most of the riverstone of steep scarp denoting cliffing positions mark the undercutting of hard rocks and thus mechanical collapse. Such collapse usually takes place along a curved, concave downwards, collapse face and the fallen block commonly disintegrates into various parts due to impact of its falls. The mechanical collapse causes due to different factors like wind undercutting, water seepage removal or a clay band beneath a sandstone solution caving and exploitation caving etc. The limestone caves also grow by this process. The mechanical collapse can be seen near Bansachua spring conditions, Manikpur, Manikpur hills near Sati Anusuiya Ashram, the left flank of river Paisuni bearing steep scarps of Vindhyan plateau, Gupta Godavari Cave I, Gupta Godavari Cave II etc. The Cave position is formed by the interaction of solution while the other examples found in sandstone country are existed by water seepage and river actions of that place.

On the basis of the above discussions of physical weathering processes, it may be inferred that the thickly forested areas of the southern upland are less disturbed by physical weathering processes. Similarly the top surfaces of the plateau are slightly influenced. But the thinly vegetated barren land and several residual hills mark intense physical weathering.

The physical and geochemical processes of disintegration, decomposition transportation and redeposition of mineral materials are termed as supergenesis. Thus the physical weathering is an act of supergenesis of rock mass of the area.

CHEMICAL WEATHERING

The study area experiences more or less all the important phenomenon of moderate chemical weathering such as high temperature throughout the year except winter months, high rainfall during the rainy months (June, July, August and September) sufficient relative humidity, infiltration of rain water (15 to 25%) and thick litter of leaves. Chemical weathering has a deeper penetrative capacity and its impact is defined by a radical transformation of rocks. This type of weathering consists the effect of various atmospheric factors; like oxygen, carbon dioxide, air moisture and the active organic substances produced by the vital activity of plant and animal organisms (Prasad, G. 1987). The major reactions causing chemical weathering are as oxidation, hydration, dissolution and hydrolysis.

It is observed that the intercepted rain water due to the forest canopy does not directly reach the ground surface but follows the route directed by leaves, branches and stems of trees and thus as a 'aerial streamlets' it infiltrates in the eluviation zones and gradually moves downward in the illuviation zones. It generates thick regolith cover. The top surface of the plateau generally where the sandstone and limestone formations are horizontally bedded, the rain drops directly reach the ground surface due to less vegetal cover and along the pores and spaces of sandstone rocks it dissolves the soft clayey materials in the eluviation zone. The dissolved materials move downward making several pits by water-born solution actions. Solution action is observed over the sloping ground of Nilgarh Pahar near Bharatkup. The regolith is fully covered with bushes. The top surface of Gupta-Godavari hill also marks solution due to several occasionally streamlets. The aerial streamlets generally weathered the loose sandy-clay materials and thus a weathered soil profile may exist. A weathered soil profile is noticed over Argahuva Pahar near Bharatpur. It is caused by solution and material transportation.

Oxygen presence in the volume of water is capable to converting many minerals into oxides and thus oxides further converted into hydroxides. This process is termed as oxidation wherein the weathering residue becomes in red coloured due to the formation of iron oxide. It is observed that most of the eroded materials of the study region are found in red colour because of ferrous and sulphides mineral concentration where the ferrous oxide immediately liberates from the rock mass. It oxidises into ferric oxide ($4\ FeO+O_2 \rightarrow 2\ Fe_2O_2$) and iron sulphite changes into sulphate of iron ($FeS+2O_2 \rightarrow FeSO_2$) and thus it generate reddish colour of loose materials.

Corbonation process is more effective in the region. The thickly forested localities of the plateau highly produces the carbon dioxide (CO_2) for corresponding soil horizons and when the rain water besides atmospheric carbon dioxide corresponds these soil horizons, it forms carbonic acid ($H_2O+CO_2=H_2O_2$). Thus the carbonic acid or carbonated water attacks the rocks and minerals and bring them into solution. Most of the formation of limestone in the area have been dissolved by carbonation process ($CaCO_2+H_2CO_3 \rightarrow Ca\ (HCO_2)_2$.

The incidence of thunder storms and associated rain fall during the rainy months produces nitric acids of helps to rain water for chemical reactions. Sometimes organic acids and nitric acids act together but both the acids in water and their effect on the decomposition of rocks are more or less similar.

Most of the granular disintegration of the granite occurs along the several isolated hillocks, are due to the chemical reactions of mineral and water. The process is termed as *hydrolysis*.

The larger area experiences *eluviation* or *leaching* which in turn initiates *rainwash* or *humification* process. *Creep* or *solifluction* are the other important microprocesses. The pH value is measured more than 7 which indicates that the sandstone is a calcite dominated and largely affected by solution actions. The initial porosity of the rock (18%) and high relative permeability (500) encourage solution action to a greater depth. The different pores, spaces, fissures, joints and bedding planes of Kaimur sandstones carry much variation in water holding capacity and thus indicate the larger degree of geophysical and chemical reaction.

Apart from the above examples of chemical weathering, there are several cases of rock decomposition in different localities of sandstone and limestone structure which are exposed in the river beds in various size and shape. Several micro-level land forms like circular and elliptical pits are formed over the dolomite structure due to solution action of running water near Sitakund, Chitrakut. Another example of 'Shiprock' found in Paisuni river bed near Sati Anusuiya indicates chemical weathering of sandstone rock due to water solution. The formation of Gupta-Godavari caves is also defined by solution. Several scratches, micro level pits and cavity are as the result of chemical weathering. The ravination processes have caused the middle course of waterbodies into ravinous land.

It is necessary to note that the chemical weathering and its measurement is a typical job for geomorphologists so that the reactions assessment are made on the basis of geomorphological principles.

BIOTIC WEATHERING

Plants, animals and bacteria largely control the breakdown of rocks and minerals. Polynov, B. (1937) believes that completely sterile weathering is impossible. Vegetation litter and decaying vegetation—various stages of humus—are important in helping to conserve moisture which in turn enhances weathering. Another important effect of vegetation is the formation of 'leaf leachates' which are very important in cheluviation. The tree leaves actively mobilise iron than are grasses.

Leaf leachates can speed up the process of pedozolisation to an extraordinary degree and different species of tree have different effects in this respect. Some trees that appear from experiments on their leaves, to be quite capable of producing a podzol are in fact never found on pedzolic soils. Because all the plants tested in the laboratory give extracts capable of mobilising iron, regardless of their behaviour in the field, factors such as the proneness of the phenolic compounds to suffer oxidation and the palatability of the litter to soil fauna may ultimately determine whether pedzolisation will occur under a given species (Ollier, C. 1969). The vegetation controls the erosion and its important role is to ease the removal of watershed products. Walker, E.H. (1963) described the relative rates of erosion under grass and forest. He advocated that the erosion is at least twice as fast under grass and it is most likely that weathering therefore also be faster.

The dominant role of vegetation on weathering of rocks is very much important; physically and chemically. Small plants like grasses, bushes and thorny scrups which are most common over the isolated hills, ravine zones and on the plateau country are more or less moisture loving. The rocks upon which they took their birth has kept continually moist. The moisture or acidic humus acts in various ways in decomposing and disintegrating the rocks. Most of the tree grains entrance into small cracks and crevices of the rocks continue to grow in chinks. Growing roots exert strong pressure on the walls of cracks and acting as wedges pry them apart into slabs and fragments. The dead roots of plants become swollen with rain water and also widen the cracks. They give rise to certain organic compounds (mainly an acid type) during decomposition which attack the associated rocks and dissolve some of their constituents.

Some peculiar examples of biotic weathering can be quoted from the different parts of the study region. Disintegration of rocks due to plant roots is observed near Manikpur hills. Another example has obtained from the alluvium deposits near Sitapur where tree roots are safe-guarding the soil profile. Near Phatikshila, the tree roots have caused a small weathering pits. The dry bed of Gupta-Godavari river and the vertical wall of Gupta-Godavari Cave I indicate weathering by plants.

Human interferences have also caused the mechanical weathering. Man's technological activities such as engineering, mining and agriculture have greatly encouraged to weathering processes. Human interferences are seen along Manikpur-Satna railway line where boulders are breaked by bomb blasting to construct new railway line and near Bharatkup, along the eastern flank of Argarahua pahar.

The effect of animals, worms and insects on weathering is also notable in the present study region.

MASS MOVEMENT AND ASSOCIATED PROCESSES

The term 'mass movement' is applied to those processes which involve a transfer of slope forming materials from higher to lower ground, under the influence of gravity, without the primary assistance of a fluid transporting agent. They merge imperceptibly with processes in which transporting media, such as air, water or ice are involved and which are generally called 'mass-transport processes'. The movements may be slow or rapid, shallow or deep and include one or more of the mechanism of creep, flow side or fall (Brunsden, D 1979).

Mass movement occurs when the disturbing forces become greater than the resistance of the slope forming materials. The movement of dry rock or weathered material is one extreme example and the flow of water containing a few sedimentary particles is another extreme case and between these two extremes is a range of mass movement types which could be classified according to the relative proportions of sediment and water. Mass movement can be based upon the three modes of movement namely, flow, slide or subsidence/heave. Hutchinson, J.N. (1968) has distinguished four types of mass movement phenomena namely, creep, forzen ground phenomena, landslides and subsidence. The following mass movement processes can be traced in the present study region.

Landslides

Landslides are relatively rapid movements of slope-forming materials in which failure takes place on one or more discrete surfaces that limit and define the failed mass. Sliding failures may occur in soils, as shallow planar slips, as debris slides, as rotational, deep-seated mass movements, as mud slides or in complex forms. The movement may be translational, rotating or retrogressive and in single, multiple or successive arrangements. Slides affect both hard rocks and unconsolidated rejolith material. The moving mass slides down an inclined plane more or less intact until it reaches the bottom of the plane where impact usually breaks it up (Clowes, A. and Comfort, P. 1982). The landslides cause by external and internal changes of shearing resistance without any change in the shear stresses. External changes break the stability conditions of rocks causing by geometrical changes, unloading, loading, shocks and vibrations, liquefaction, remounding, fluidization, air lubrication, cohesionless grain flow and changes in water regime. Internal changes occur due to progressive failure, weathering and seepage erosion.

A massive rock slide can be seen near Sati Anusuiya Ashram where the river Paisuni has deepened the valley making steep scarp. It is found that the scarp zone generally shows rock slide due to disintegration and sliding of rock mass.

Rock-Falls

Falls of rock or soil involves the free movement of material away from a steep slope. The size and shape of the detached mass is usually dependent on the nature of the discontinuities in the rock, the state of weathering of the rock mass and the slope geometry. Rock falls are always derived from the superficial layers of the rock face, a feature that distinguishes them from rock slides or rock avalanches which may be deep-seated (Brunsden, D. 1979).

The steepest slopes of 70°-90° usually occur falls where the angle of friction is greatly exceeded. In rock fall, the detached particles move down the slope by a combination of falling, rolling and bouncing until it reaches a point where the slope angle is low enough to allow the particle to come to rest. Rock falls also cause either by climatic variables and surface weathering or in the case of larger rock falls due to undercutting by erosion unloading, lateral expansion, the development of tension cracks and stress-concentration processes. High rail fall, freeze thaw and desiccation weathering cause the small rock falls.

This problem originates in the work of Terzaghi, K. (1962a), Skempton and La Rochelle (1965), Bjerrum and Jorstad (1968), Gardner, J. (1970a), Hutchinson, J.N. (1971a) etc. In the case of weathering oriented rock falls, it adopt distinctive modes of initial failure when larger masses are involved. These are mainly characteristics of well jointed, bedded materials and take place along joints or deep tension cracks. Plane, wedge and slab modes of failure are the most common. Several examples of rock falls can be quoted from the region. It can also be expressed that the dissected hill side slopes near Sheorampur have loosened the sandstone boulders between the loose merrum deposits Gupta-Godavari caves of limestone formation, indicates the collapse of rock masses due to solution and weathering processes.

Slumps

Slumps usually occur in weaker rocks than slides. It is distinguished from them by having a rotational movement along a curved slip plane. Small scale slumps occur on debris-covered slopes which have a vegetation cover.

Slumping can be observed near Sheorampur hills where loose materials slump downward due to weathering and occasional heavy rainfall. Khoh Pahari, about 8 km away from Karwi along Karwi-Allahabad road show minor slumps. Ravinous wall of Yamuna river (near Rajapur) also indicates few peculiar examples of slumps. Slope materials more than 35 per cent by sedimented loads are prone to movement by flow. It is the movement of mass by internal deformation under its own weight. Sheorampur hill, Nilgarh Pahar, Neolakhoh Pahari and area along the Yamuna river near Rajapur show some interesting examples of flow marked in the various localities of the region.

Surface Wash

Surface wash is not a mass movement phenomenon. It occurs on slopes when either the intensity of rainfall exceeds the capacity of the soil to absorb water, the infiltration capacity or where the water table lies at the surface. Overland flow causes the surface wash. Sheet flow and raindrop action are the factor by overland flow during surface wash. It is remarkable that weathering accelerates the surface wash. Raindrop action causes sheeting in most of the area away from the river catchment.

CONCLUSION

On the basis of the above discussion, it may be pertinent to point out that mass movement is the process which exists due to weathering determinants and running water. It is also remarkable that landforms are the result of the complex interactions of weathering and erosion. Physical and mechanical weathering have affected at the greater extent while the role of chemical and biotic weathering have also reached to greater intensity. It is impossible to check the nature of processes response because the complexity in destructive variables is insignificant.

5

Fluvial Processes

The present chapter deals with the study of fluvial processes operating the existing terrain of the region. The recognition of fluvial processes and its intensity have been considered as a difficult task for geomorphologists, therefore the study relates around the realm of secondary data, topographical outlooks and general field surveys.

The overall comments on erosional typology and techniques have been postulated on the basis of the field observations concluded by several geologists from time to time. The review of system and mechanism of fluvial processes incorporating infiltration, length of overland flow and surface flow have been made with the help of data obtained from several development blocks and selecting 25 drainage lines of the region. The interpretation of surface flow is stated through the measurement of meander symmetry and sinuosity indices of 25 drainage lines. The remaining processes like pipe flow, subsurface flow, sheet wash and sheet flow and rill wash and gullying have been discussed on the basis of the field survey.

INTRODUCTION TO EROSIONAL TYPOLOGY

The geological work of flowing surface water confirms the fluvial processes and fluvio-culture of the region. Its intensity relates to mass of water and flow velocity. As mentioned earlier, Chitrakut and its adjoining areas are the part and parcel of Vindhya mountains which existed from Vindhyan basin. The processes which are operating today are not the same as operated in the whole span of time of Vindhyan history. They differ in nature due to variability in climatic conditions. The geologists approach on stratigraphy of Chitrakut formed in different geological time also confirm the changing operational characteristics of the processes. But the up-to-date Vindhyan formation clearly suggests the dominance of fluvial processes through the passage of time. Following the scenery of present day landforms and climatic conditions, the entire region is considered as a fluvially dominated terrain.

The intensity of present day processes operating the existing terrain have been determined with the help of Peltier, L.C. (1950) and Wilson, L. (1973) models depending upon the mean annual temperature (26.02°C) and mean annual rainfall (95.65 cm) of the region (Chapter-3). In this measurement, the region falls under moderate morphogenetic characteristics where fluvial erosion is marked at maximum intensity. Thus, the present day landscapes are shaping form under the effect of fluvial processes.

But the geological history of the environs of Chitrakut depicts of a few peculiar examples of erosional typology. The observations of Kedar Narain (1960) are appreciable in this context. He stated that the Semn group (1100-1400 M. years) is characterised of calcareous facies which is also referred as to platform, stable shelf or fore land facies. The associated sandstone (910-940 M.Y.) is orthoquartzite type wherein most of the unstable minerals have been winnowed out owing to protracted and profound weathering and abrasion. The presence of few fragments of meta-quartzite and chert together with well rounded grains of quartz suggest that the orthoquartzite has passed through more than one cycle. The complete absence of felspars further indicate that a prolonged washing and winnowing is responsible for the generation of supermature sands of the region. Kedar Narain advocated that the glauconite (1110 + 60 M.Y.) of the region is formed in a marine environment at relatively shallow depths which might have reached the stage of peneplanation or base-levelling. It follows that either the rivers flowing into the basin had reached maturity and were not capable of carrying such materials or there were no major rivers flowing into it. The granular nature of the glauconite and pellet limestone of semri in the region support shallow depositional basin agitated by currents which were quite effective at the time of formation of these sediments. Possibly the sea transgression and submergence processes for a short duration were common during glauconite formation. It also supports shallower basin conditions and a process of regression after the formation of pellet limestone in the area. Dubey, V.S. and Chowdhury, M.S. (1952) suggested that the processes during the Semri formation were related to a warmer climate than to a cold climate. The presence of disintegrated and weathered pebbles and boulders of granite in Basal congromerate suggests that the chemical decomposition processes were at its lowest ebb and mechanical disintegration was the main process of weathering which is characteristic of arid zones where the precipitation was its minimum and the fluvial processes were less effective. The complete absence of felspars in the lower Glauconite sandstone indicates a change of processes from arid to humid and warmer climate. The formation and precipitation of calcareous material, oolites etc. show warmer climate and associated processes. Thus, as a whole the Semri beds of the region were formed through the processes operating under warmer climatic conditions.

Kedar Narain further suggested that the Kaimur formations in the area were laid down in the near shore zone. The presence of sun cracks, ripple marks and mud cracks in the bottom beds subjected alternatively the processes working under dry and wet conditions. In this period, a flat coast line near the mouth of large rivers carrying down sands in an easterly direction across a coastal plain. The claygalls and rounded grains suggest the succeeding flood conditions and the next high tide or storm. The formation of quartzite in the region suggests a sudden increase in the potency of winnowing and or abrasion by wave or current actions. Thus the Kaimur sedimentation is processed in an high energy beach environment where the waves could work upon these for a long time and thus effect the winnowing and or abrasion of unstable materials and rounding of the sandgrains. The Kaimur red sandstone might imply deep weathering of ferro-magnesian minerals in the source area under semi-arid conditions (Kedar Narain, 1960).

Apart from the above discussion, several Geologists have also discoursed the nature of the processes in Vindhyan period. Roy, A.K. and Bhattacharya, A. (1982) reviewed that the drainage patterns were mostly dendritic, subdendritic and joint controlled. They suggest that the older granitic pediplain represents an old erosion surface and the three planar surface with an average elevatiòn of 100 m, 150 m and 300 m in Bundelkhand pediplain province indicate successive erosional activities in a cyclic form. They further explained that palaeo erosional surfaces or exhumed surfaces of Vindhyan landscape are caused due to the process of exhuming which is still in operation.

Valdiya, K.S. (1982) suggests that in the Palaeozoic period the rivers flowing north-westerly and westerly directions deposited thick continental Gondwana sediments. Ravi Prakash and Delela, I.K. (1982) also recognised a mature erosion surface underlying dominant phase of sandstone. The act of deformation and erosion is caused due to uplift of granitised blocks during pre-Cambrian granitisation in the region.

Interesting results regarding the history of the operating processes appear on the basis of the analysis of the paleocurrent data. It is thought that the Vindhyan basin has shown the dominance of a unimodal north-westerly paleocurrent pattern (Banerjee, I.B, and Sengupta, S. 1963, Banerjee, I.B. 1964, Mishra, R.C. 1964 and Banerjee, I.B. 1974). On the open shelf in the northeast part of the basin, tidal currents would be shore normal landward and seaward and therefore biopolar on the simplest assumptions. However, due to the coriolis effects tidal currents in the open shelf follow elliptical paths with flood currents veering to the right in the northern hemisphere and the ebb current to the left (Defant, A. 1958). In such a system, the tidal waves move in a rotary anticlockwise path around the point of constant sea level. Tidal currents at any point are rectilinear but show a regionally complex pattern. In the Vindhyan sea the surface and bottom currents responsible for sediment movement are dominantly: (i) wind driven; (ii) wave-induced; and (iii) tidal in origin (Banerjee, I. 1982). Storm surges add to normal wave agitation and seaward-returning storm surge ebb current is a dominant agent of deposition in the nearshore environment (Kummar, N. and Sanders, J.E. 1976).

Chanda, S.K. and Bhattarcharya, A. (1982) analysed that the barrier bar complexes of Vindhyan basin were sub-merged during rising sea level and reworked *in situ* by shelf tidal currents into subtidal sand-waves. Alternately shoreward development of fine grained (lagoonal) facies may also owe its origin to weakening of ocean and tidal currents in the sand-wave field further offshore. Notwithstanding a wide spectrum of shallow marine environments, regionally persistent northwesterly paleo-current suggests that tidal currents were definitely reinforced by another set of current most probably wind drift current. These paleo-currents were the main root cause of erosional and depositional activities in the Vindhyan basin.

Singh, I.B. (1980) visualised that in the absence of stabilization by vegetation in the Proterozoic, the barrier bar complex would undergo extensive erosion and remoulding over the legoonal deposit. Near complete loss of barrier beach complex, he further suggests that transgression occurred by shoreface erosion rather than by rapid sea level rise (Reinson, G.E. 1976).

Prasad, B. (1984) described that there appears to be a quick inflow of water towards the end of deposition of Rewas. This inflow of water was for a very short duration. He further advocated that the present character of Tirohan limestone may be due to penecontemporaneous deformation and or erosion, faulting and or collapse.

Overall discussion reveals the fact that the stratigraphic sequence of Chitrakut plateau has undergone in a changing marine conditions by active paleocurrents with a significant swift in climatic regime. Since the existence of the mountain, fluvial processes accelerate their geological works and a thick amount of alluvium deposited in the northern plain by Paisuni and its major tributaries. It is notable that Vindhyan formation denotes dominance of marine processes while landscape after mountain building experiences fluvial processes. Present day erosional activity differs from basin to basin due to mass and velocity of water and from one litho-structure to another due to nature of composition of rocks. Thus, sheeting, piping, rilling, gullying etc. Processes exist at different dimension following the riverine society.

EROSIONAL TECHNIQUES

In the previous discussion, geologists approach impressed the views that the geological history of Chitrakut as well as

Vindhyan stratigraphy have experienced climatic crux in phases. The marine environment as agitated by turbulent paleocurrents was thought to be warmer, humid, arid and semiarid in nature where regression, transgression, exhumation, denudation, winnowing, abrasion, shifting of land and sea and emergence and submergence of sea level have technically processed the suits of stratigraphical sequence with the march of geo-chronology. These techniques imprisoned in basin structural history undoubtedly at the time of mountain building of Vindhyachal as well as Chitrakut itself. Since existence, the uplifted landmass has undergone several tectonic jerks and still stand has faced the impact of exogenetic processes. The denudational processes and several other erosional techniques cycled the plateau in senile stage. The transporting agencies (river-bodies) have made more complex terrain profile plan controlling with erosional, transportational and depositional techniques. The presentday assemblage of landforms recite the above fact.

The records of geologists observations highlight the formational and deformational techniques of the mountains. Roy, A.K. and Bhattacharya, A. (1982) opined that due to gradual uplift and change in palaeogeomorphic set up the present topography has evolved. The uplifted topography was later subjected to process of denudation, scarp retreat and peneplanation giving rise to the platform. Quite later, the Vindhyan formations and topography were covered under profuse outpourings of the Deccan lavas. Denudational techniques have exhumed the Vindhyan topography which continues backwearing of scarps even to the presentday. Between late pre-Cambrian to early Cambrian, it was uplifted and peneplained in several phases since then. The evidence of peneplanation is afforded by the presence of many flat-topped hills at different levels with the same elevation. It is observed during the field survey that the scarpfaces occur vertical cliffing generally underlying rectilinear or concave slope profile in plan. The free face element of the scarp is actively backweared by parallel retreat processes. The erosional techniques as suggested by Davis i.e. slope declining is also progressed here because most of the hill ranges have diminished or lowered at a greater extent. The rectilinear slopes of the scarp which are systematically covered with erosional sediments are found in ruined form due to rilling and ravaging impact. The alluvium country specially along the river courses experiences alluvium mudflows, piping and rilling actions. Thus the erosional techniques from later geological history to presentday and from place to place are variable in nature.

DETERMINANTS OF FLUVIAL PROCESSES

The character of each river channel is determined by three principal features like:

1. the quantity of water carried by them;
2. the level of water; and
3. flow velocity.

All these factors do not remain constant but change from season to season and from year to year. The pattern of these changes is called as the regime of a water. The work of river erosion and transport requires energy. The energy of the river is provided by the velocity and volume of water. Velocity depends upon:

1. the gradient or steepness of valley. The steeper the slope, the greater is the velocity.
2. the slope of the valley or the form of the channel. Velocity will be greater in the channel with deep and narrow form than in the other with broad and shallow form.
3. the volume of water. The velocity of water increases with the increase in the volume of water and that is why during the flood period, the discharge increase rapidly.
4. the amount of load carried by a river. If a river carries a large quantity of rock materials, part of its energy is expended in the work of transport and consequently its velocity and erosive power diminish although the materials borne by it erode its bed.

Over the mountainous country, rivers have V shaped valleys and steep gradients so that their velocities are high but in the flat plain country, the volume increase while the valley becomes broad and shallow and gradient becomes gentle. Here the large cross section of the valley permits greater discharge in spite of reduced velocity.

The quantity and the level of water in the river depend on the sources feeding it and on their seasonal variation from one place to another. Surface and ground water also play a role in feeding rivers. Depending on the prominent source rivers can take melting snow water, glaciers waters and water supply mostly from rain. The nature of water mass also determines the erosional capacity.

The period of least amount of water in the river and its lowest level is known as low water time or low water level. The period when a river is characterized by a sharp increased in the amount of water and rise of its level as a result of melting of snow or heavy rain is called high water. The level of water in a channel also determines the characteristics of erosional capacity.

The rate of flow of water in a river depends on the mass of water, the gradient and the character of the channel which determines the specific features of the flow. In rivers, the flow of water is mostly turbulent i.e. the rate of flow at each point of the stream is not constant either with regard to velocity or direction.

SYSTEM AND MECHANISM OF FLUVIAL PROCESSES

The movement of water on slopes is governed by a complex set of interrelated factors. Rainfall duration and intensity exist as an external factors of fluvial system. The major controlling factors are intrinsic to the system and are determined by soil, vegetation and topographic characteristics. Water regime has a fundamental control on existing terrain and associated soil groups. It is difficult for geomorphologists to measure the mechanism of fluvial processes because it is the subject broadly defined by the disciplines of geohydrology, geophysics, geochemistry, bioscience and geomorphology. Thus, the observation is not possible without laboratory test and field measurement, while the geomorphology only provides the field knowledge. The efforts of a common geomorphologists about the mechanism of fluvial processes can be expressed in the following manner.

Infiltration

Infiltration may be simply defined as the process of water entering the soil. It is not similar to the hydraulic conductivity of soil which affects the rate of infiltration but it may be expressed as the flux of water across the soil surface. Rubin, J. (1966) has identified the rain infiltration on the basis of relative rainfall intensity. The rate of infiltration of water into the soil differs with time. Clowes, A. and Comfort, P. (1982) believes that the rate can be quite high at the onset of rain because of the dry soil. The small voids between clay and silt particles reduce the infiltration rate. In comparison sand and gravel have high infiltration rates as a result of the large voids between the particles. Thus the infiltration capacity is influenced by a number of factors like rainfall intensity, duration drop size and soil characteristics (texture, structure, depth, nature and proportion of clay minerals, vegetation and landuse).

The region receives about 1023.46 mm of annual rainfall. The average normal annual rainfall is obtained as 956.6 mm wherein degree of intensity decreases in all directions with significant increase of infiltration index (Table 5.1 and Fig. 5.1) to the outer margins of the region. Fig. 5.1 reveals the fact that the annual rainfall increase as well as relief variation increases. The plateau region (Naraini, Karwi, Manikpur, Pahari and Majhgawan blocks) experiences more rainfall than that of alluvial plain. Infiltration increases from hard rock (Kaimur sandstone and tirohan limestone) composition (15%) to sand, clay and morrum dominating alluvium layers of Trans-Yamuna plain (25%). The compact structure of Kaimur sandstone and few parts of older granite rocks (parent materials) which have small voids between the two particles over the plateau region reduce the infiltration rate. On the contrary the sedimentated loose sand, gravel, clay and morrum of northern plain having large voids between the particles have allowed high rate of infiltration.

Water transmitted between peds (soil structure) as well as particles. It is observed in plateau region that a well-developed crumb structure has numerous well-connected voids which have allowed the water movement in all directions. In the case of blocky structure having regular small size voids, the water moves in all directions. Prismatic structures possess large peds with well defined voids having a dominant downward orientation and water movement in vertical directions. Platy structures are seen having generally large peds with ill-defined voids possessing a dominant lateral water movement. The development of crusting, surface erosion, water repellency, accessions of salt in the profile at the end of summer season and specially freezing of water in the sub-surface, all cause wide temporal variations in the rate of infiltration. Vegetation covers also causes a considerable amount of seasonal variability. The presence of vegetation tends actually to encourage infiltration by inducing high soil water deficits.

Table 5.1: Rainfall and infiltration index

S. No.	*Name of the development blocks of Banda and Satna districts*	*Normal annual rainfall in mm*	*Rainfall recharge as % of monsoon rainfall*	*Infiltration (%) rate*
1.	Naraini	997.7	16.98	15
2.	Karwi	1017.7	11.96	15
3.	Manikpur	1017.7	12.92	15
4.	Pahari	1017.7	11.23	15
5.	Ram Nagar	947.1	38.99	15
6.	Majhgawan (Satna)	1145.0	11.85	15
7.	Barokhar	887.1	34.25	15
8.	Tindawari	887.1	33.91	25
9.	Jashpura	887.1	20.65	25
10.	Mahuwa	997.7	32.10	25
11.	Mau	947.1	16.50	15
12.	Baberu	881.2	11.41	25
13.	Kamasin	881.2	36.69	25
14.	Bisenda	881.2	30.78	25
	Average	956.6	22.87	19.29

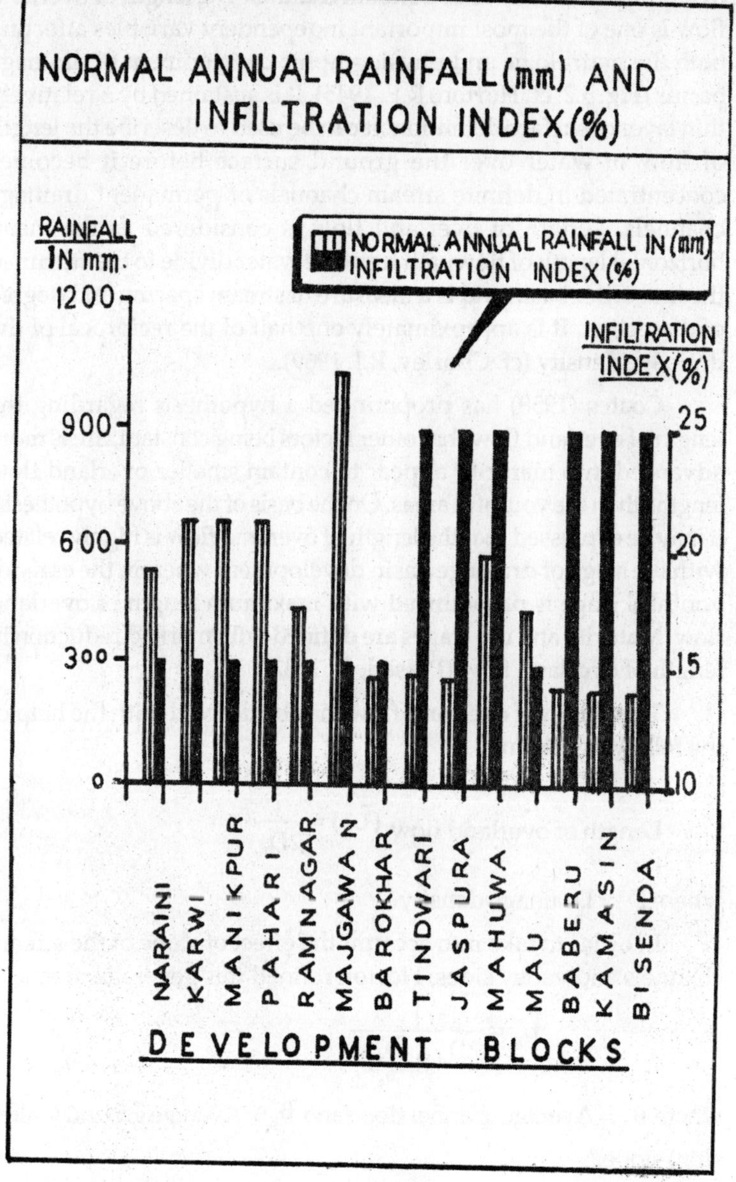

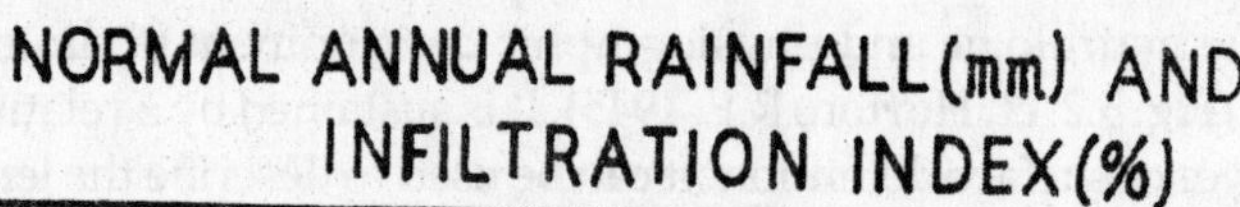
NORMAL ANNUAL RAINFALL(mm) AND
INFILTRATION INDEX(%)
RAINFALL
IN mm.
NORMAL ANNUAL RAINFALL IN (mm)
INFILTRATION INDEX (%)
INFILTRATION
INDEX(%)
1200
900
600
300
0
25
20
15
10
NARAINI
KARWI
MANIKPUR
PAHARI
RAMNAGAR
MAJGAWAN
BAROKHAR
TINDWARI
JASPURA
MAHUWA
MAU
BABERU
KAMASIN
BISENDA
DEVELOPMENT BLOCKS

Fig. 5.1

Overland Flow

Overland flow is that part of the lateral inflow which flows overland surface towards a stream channel. The length of overland flow is one of the most important independent variables affecting both the hydrologic and physiographic development of drainage basins (Fig. 5.2. cf. Horton, R.E. 1945). It is sustained by a relatively thin layer of surface detention. It can be used to describe the length of flow of water over the ground surface before it becomes concentrated in definite stream channels or permanent drainage channels. Length of overland flow is considered as the mean horizontal length of flowpath from the water divide to the stream in the first order basin and is a measure of stream spacing and degree of dissection. It is approximately one-half of the reciprocal of the drainage density (cf. Chorley, R.J. 1969).

Coates (1958) has propounded a hypothesis regarding the length of overland flow that other factors being constant, areas more advanced into maturity appear to contain smaller overland flow lengths than the youthful areas. On the basis of the above hypothesis, it may be expressed that the length of overland flow is highly related with the stage of drainage basin development wherein the early or youthful stage is pronounced with maximum length of overland flow. Maturity and old stages are defined with marked reduction in length of overland flow (Prasad, G. 1987).

The length of overland flow may be derived with the help of the following formula:

$$\text{Length of overland flow } (\bar{L}_g) = \frac{1}{2D_d}$$

where D_d = Drainage density.

In order to take into account the effect of slope of the stream channels and valley sides, Horton refined this generalization as:

$$\bar{L}_g = \frac{1}{2D_d\sqrt{1-(\frac{\theta_c}{\theta_g})^2}}$$

where θ_c = Average channel slope and θ_g = Average ground (valley side) slope.

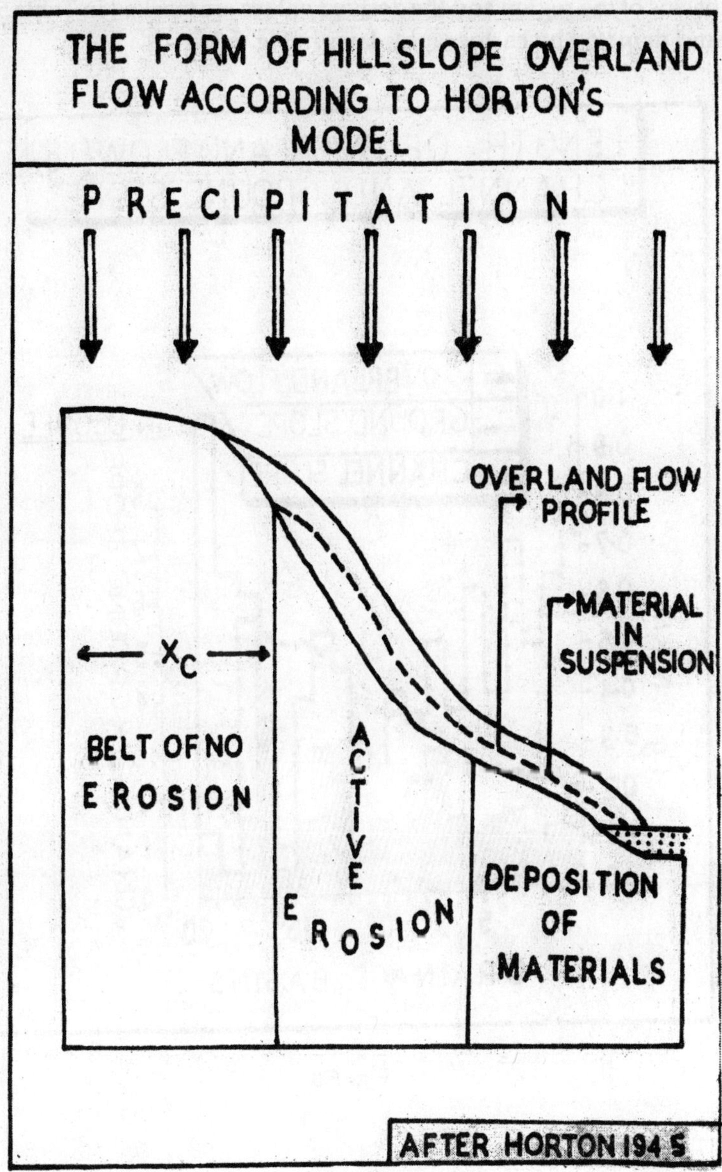
THE FORM OF HILLSLOPE OVERLAND FLOW ACCORDING TO HORTON'S MODEL
PRECIPITATION
OVERLAND FLOW PROFILE
MATERIAL IN SUSPENSION
X_C
BELT OF NO EROSION
ACTIVE EROSION
DEPOSITION OF MATERIALS
AFTER HORTON 1945

Fig. 5.2

Horton's above formula has been applied in the present case. The length of overland flow ($\bar{L}_g$) have been obtained for 20 drainage basins of the region and the derived values are marked in Table 5.2 and depicted by cartographic laws in Fig. 5.3.

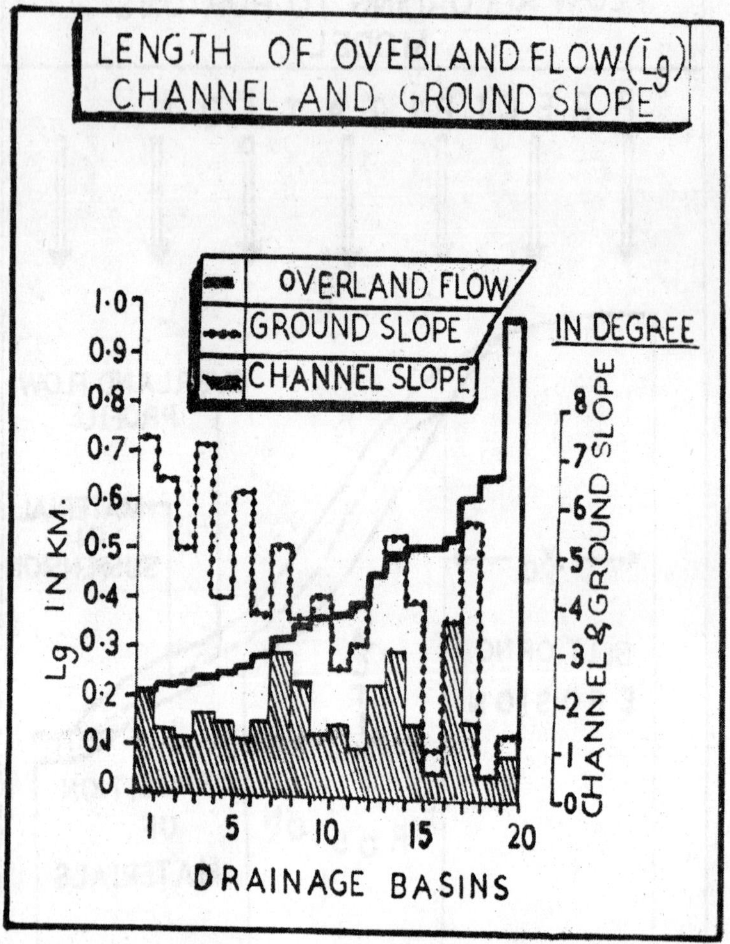

Fig. 5.3

The derived values of overland flow vary between 0.2 (Barwa basin) and 0.99 (Samrani basin). The observation closely corresponds the physiography of the region. Samrani (0.99) and Kachchua (0.67) draining over Yamuna plain, show higher values of overland flow and indicate the late mature and old stage of basin (terrain) development. Basins like Paroi (0.64), Dhora (0.51), Gauhia (0.51), Dingchi (0.51), Piprawal (0.45) and Jhurai (0.40) which are mostly developed on flat and rolling surfaces except their upper reaches also contain greater the length of overland flow. Moderate length of overland flow is observed in the case of Kewai (0.24) and Barar (0.33) whose upper and lower catchment areas are found over highly dissected hilly terrain of Chitrakut plateau and flattish terrain of Yamuna plain respectively. Barwa (0.20), Ghinhal (0.27), Jhuri (0.23), Bagdara (0.36), Kalibarah (0.22), Maro (0.28), Sukra (0.35), Kulagara (0.25) and Pindrah (0.35) basins receive minimum values due to hilly terrain having steeper the valley side and ground slope. These basins are of late youth to early mature stage. Ganta (0.54), Pindrah (0.37), Sukra (0.35) and Bagdara (0.36) basins register higher values while drain over hilly terrain. It is because of moderate slope angles along the channel and valley side and compact lithological controls of Kaimur sandstone.

On the basis of the above discussion, the following concepts may be suggested:

1. The length of overland flow increases as well as the channel and ground slopes decrease over the homogenous lithological controls of the region.

The above hypothesis holds good in the case of Samrani and Kachchua drainage basins.

2. Lithological heterogeneity and steeper the channel and ground slopes cause lower the length of overland flow and much drainage dissection mostly over hilly terrain.

   It is found correct in the proceeding discussion.

3. Higher the values of length of overland flow older the stage of basin development while lower the values of $\bar{L}_g$ indicate early stage of terrain development.

Table 5.2: Channel and ground slope and length of overland flow

S. No.	*Drainage basins*	*Channel slope (θ_c) in degree*	*Ground slope (θ_g) in degree*	*Length of overland flow (in km)*
1.	Samrani	1°	1° 12′	0.99
2.	Kachchua	30′	30′	0.67
3.	Paroi	1° 30′	5° 58′	0.64
4.	Ganta	3° 24′	3° 48′	0.54
5.	Gauhia	30′	1°	0.51
6.	Dhora	1° 36′	4°	0.51
7.	Dingchi	3°	5° 26′	0.51
8.	Piprawal	2° 20′	4° 36′	0.45
9.	Jhurai	1°	3°	0.40
10.	Pindrah	1° 24′	2° 8′	0.37
11.	Bagdara	1° 18′	4°	0.36
12.	Sukra	2° 18′	3° 36′	0.35
13.	Barar	3°	5° 5′	0.33
14.	Maro	1° 30′	3° 45′	0.28
15.	Ghinhal	1° 12′	6° 10′	0.27
16.	Kulagara	1° 30′	4° 4′	0.25
17.	Kewai	1° 36′	7° 6′	0.24
18.	Jhuri	1° 15′	5° 2′	0.23
19.	Kalibarah	1° 20′	6° 28′	0.22
20.	Barwa	2°	7° 16′	0.20

Note: The serial numbers of drainage basins have been changed and arranged in the descending order of the values of length of overland flow.

The results of 20 drainage basins validate the above concept. But the rule contradicts the valuable finding of Coats (1958) who related the values of $\bar{L}_g$ with the stage of basin development. The postulation of Coates is not tenable because the $\bar{L}_g$ is directly negatively correlated with drainage density and highly affected by geological structure, lithology, slopes and vegetational cover etc. of a region. It generally decreases with increasing drainage density. On an average drainage basin develops maximum stream segments in its late youth and early mature stage hence minimum length of overland flow must be found.

Thick sod cover is expected to be the most important controlling factor of length of overland flow besides the rainfall intensity and infiltration rates. The catchment area of several drainage basins of northern plain are covered with thin vegetational properties and exhibit longer the length of overland flow. The crest of the scarps (flat-topped surfaces) are also characterized by the longer overland flow paths because of the uneroded water divides having less vegetal properties.

Concentrated tshrough Flow or Pipe Flow

Pipe flow occurs generally by those soils which have either peaty surface horizons and impermeable layers at shallow depth or those where surface horizons are loamy and the slope is steep. Piping has been recorded in many areas of the world. It provides a macropore network for the quick transmission of through flow. Jones, J.A.A. (1971) has marked several interesting areas of piping in the British Isles. Gilman, K. and Newson, M.D. (1980) have also investigated few examples of pipe networks on upland wales.

Eluviation and desiccation are the most important processes which produce pipes. Eluviation allows the fine fraction to be removed progressively through the soil matrix. This is indication by the deposition of the clay and silt where the pipe emerge. But, eluviation sufficient to lead to the creation of pipes only occurs in soils possessing an initial weakness caused by low bulk density, an unusual particle size distribution or a structure which has been altered by chemical effects. The aggregation-dispersion status of the soil seems to be important (Gerrard, 1981). Some workers have argued that pre-existing voids are necessary for the development of pipes.

A wide literature on the creation of pipes by desiccation is available. The work of Henkel, J.S.A.W. (1938), Heede, B.H. (1971), Huges, P.J. (1972) and Rathjens, C. (1973) on desiccation is appreciable. It is thought that the formation of pipe is still uncertain because desiccation cracking is a very complicated process depending on a number of soil properties such as strength and shrinkage potential. The types and presence of clay minerals in the soil largely affect the desiccation cracks. Parker, G.G. (1964) considered that the enlargement of desiccation cracks was the principal mechanism of piping in the cohesive clay soils whilst eluviation was important in cohesionless materials. Piping appears to be particularly prevalent in semi-arid regions where the regolith contains clays such as montmorillonite which crack on drying. Desiccation cracking is also thought to be the initiating mechanism in Upland Wales (Gilman, K. and Newson, M.D. 1980).

Piping is thought to be responsible for the movement of silt and clay particles downslope in areas where surface processes are mainly ineffectual. In the present study, the pipes are measured along the river banks. The desiccation cracks are seen throughout the Banda plain. It will be thoroughly discussed in Chapter 9. Piping and desiccation cracks are seen along the river side and hill side slopes of the region.

Surface and Sub-surface Flow

Water moves over the ground surface across slopes in a variety of ways. The processes like rainplash, surface wash, solution and mass movement exist, by flowing water. Wind action is also considered as surface processes. The relative importance of these processes and their interactions are largely but not entirely governed by climate. Schumm, S.A. (1956a and b, 1964) has suggested that the interaction of the various processes in any specific site can be extremely complicated. It is measured that microtopography influences the type of surface flow that takes place. Unconcentrated flow in thin sheets is only possible on fairly smooth surfaces. Emmett, W.M. (1970) found that sheet flow or wash varied greatly in character and depth and was a mixture of laminar and turbulent flow.

Clowes, A. and Comfort, P. (1982) express that the flowing water participate along an open channel does not follow a direct straightline or even a smoothly curving path or trajectory. It moves vertically and laterally in addition to its overall downstream direction. The actual velocity of this particle is recognised much greater than the mean velocity in a downstream direction. This type of flow is called turbulent flow and predominates is most natural river channels. A second type of flow is seminar flow exists by the linear trajectories of water particles. It is generally found at lower flow velocities and is of very limited occurrence in natural river channels. It is usually found that the fastest flow of water occurs in the centre of the channel. The edges of the channel show slowest surface flow. The small channels or tributaries show intense flow. The formation of rills along hill sides and river bed divert the surface flow into many channels. This type of examples can be seen along the rectilinear slopes of the plateau region. The upper flow paths of the channel are also considered with fastest flow. Similarly paisuni flow path during the rainy months exhibits the fastest flow of the time. A meandering path having a slow movement of water in a Paisuni river can be seen from the site of Yamuna-Paisuni confluence near Rajapur during the month of June. It is also remarkable that the surface flow is well governed with seasonal rainfall variations. Mostly the small tributaries of the region have no water during the summer season.

The topographic variations, lithological changes and the climatic uncertainty of the region have caused so many surface flow paths (fingertrip tributaries) along the slope dynamics. The drainage network of the region and its orientation truly explain the surface flow complexity and its orientation. It is also well recognised that the upland region denotes quick run-off for a short period of time while the lower sections of alluvial plain indicate maximum quantity of flowing water generally for a calender year but it denotes lesser numbers of surface tributaries. In order to measure the surface flow regularity and the variability raised due to different environmental factors, it has been attempted to find out the morphometric results of various sinuosity indices and determinants of meander properties for 25 drainage lines of the region (Prasad, G. and Devi, V. 1987). The measurement of meander properties includes symmetry of arc meander length, arc meander height, symmetry of form ratio and repetition of symmetry. These variables are derived with the help of the following formulae introduced by Brice, J.C. (1960, 1964):

1. Symmetry of arc meander length

$$= 100 - \frac{100\ (\text{mean deviation of arc length})}{\text{mean arc length}}\ \%$$

2. Symmetry of arc meander height

$$= 100 - \frac{100\ (\text{mean deviation of arc height})}{\text{mean are height}}\ \%$$

3. Symmetry of form ratio

$$100 - \frac{100\ (\text{mean deviation of form ratio})}{\text{mean form ratio}}\ \%$$

4. Repetition of symmetry $= \frac{N_a}{\sum N_a} \times 100$

where N_a = number of arc whose length does not deviate more than 25% from the average length of meander,

ΣN_a = total number of meanders.

Here, it is notable that the values of different components of meander symmetry, if differ between 50 per cent, the surface flow line may be defined as moderately meandering. Above 80 per cent it will denote fully meandering flow plath. In case of sinuosity index, the following formulae as suggested by Schumm, S.A. (1956) and Muller, F. (1968) have been used:

1. $C_I = \frac{C_L}{A_L}$

2. $V_I = \frac{C_L}{V_L}$

3. $S_{SI} = \frac{C_L}{V_L}$

4. H_{SI} = % equivalent of $\frac{C_I - V_I}{C_I - 1}$

5. T_{SI} = % equivalent of $\frac{V_I - 1}{C_I - 1}$

where C_I = 'Channel Index'

C_L = 'Channel Length'

A_L = 'Air Length' between the source and mouth of the river

V_I = Valley Index

VL = Valley Length

S_{SI} = Standard Sinuosity Index

H_{SI} = Hydrological Sinuosity Index; and

T_{SI} = Topographic Sinuosity Index

According to Schumm, S.A. (1963) postulation, the surface flow having the value of 1.0 is straight, just above the unity to 1.3 is sinuous and more than 1.3 is meandering.

The derived values of above two variables have been arranged in Table 5.3 and 5.4. Following to Table 5.3 and the rules regarding the meander properties of surface flow path, it is pertinent to point out that the drainage lines like Piprawal, Barwa, Maro, Ganta, Dingchi, Paroi, Jhuri, Kachchua and Geduwa show moderately meandering surface flow course. Another rivers are of sinuous in nature. In the case of repetition of symmetry, its higher percentage (50%) indicate meandering course while the lower percentage are related to straight or sinuous path of surface flow. It is notable that none of the basis has 50 per cent of repetition of symmetry and remark sinuous characteristics. But the field measurement confirms that the basins which have above 30 per cent of repetition of symmetry are engaged in making meandering flow course.

Table 5.3: Meander symmetry

S. No.	*Drainage basins*	*Symmetry of arc meander length (%)*	*Symmetry of meander width (%)*	*Symmetry of form ratio (%)*	*Repetition of symmetry (%)*
1.	Samrani	35.50	58.21	45.26	31.78
2.	Paroi	45.45	52.94	56.87	30.43
3.	Piprawal	55.42	62.90	59.89	36.59
4.	Barwa	59.64	55.00	57.85	47.50
5.	Ghinhal	47.11	43.96	63.21	25.71
6.	Jhuri	44.17	55.32	64.23	23.23
7.	Bagdara	57.62	37.63	41.19	30.43
8.	Kalibarah	36.97	66.04	31.18	25.00
9.	Maro	56.41	55.68	65.36	45.16
10.	Sukra	30.77	50.00	54.00	29.73
11.	Kulagara	32.14	65.31	50.66	21.88
12.	Pindrah	37.78	38.41	66.11	14.29
13.	Barar	37.63	65.00	48.72	10.45
14.	Kewai	25.93	42.42	36.29	17.19
15.	Ganta	55.45	61.86	52.80	20.00
16.	Jhurai	42.24	58.21	45.13	24.32
17.	Gauhia	17.95	36.76	44.25	19.23
18.	Kachchua	50.00	46.15	51.87	40.48
19.	Dhora	41.03	42.67	47.60	29.03
20.	Dingchi	58.70	51.43	55.21	31.25
21.	Jaiwanti	43.18	70.00	50.66	21.54
22.	Ganta (main)	51.47	48.28	39.39	37.84
23.	Ohan	48.48	48.31	51.56	37.50
24.	Geduwa	48.28	55.22	52.59	27.27
25.	Paisuni	51.79	56.11	49.75	32.26

Table 5.4: Sinuosity indices

S. No.	*Drainage basins*	*CI*	*VI*	H_{SI} (%)	T_{SI} (%)	*SSI*
1.	Samrani	1.63	1.56	11.67	88.33	1.04
2.	Paroi	1.42	1.36	15.17	84.83	1.05
3.	Piprawal	1.32	1.28	12.07	87.93	1.03
4.	Barwa	1.43	1.36	16.26	83.74	1.05
5.	Ghinhal	2.07	2.02	5.88	94.12	1.03
6.	Jhuri	1.31	1.27	12.13	87.87	1.03
7.	Bagdara	1.68	1.62	9.22	90.78	1.04
8.	Kalibarah	1.23	1.18	21.65	78.35	1.04
9.	Maro	1.67	1.60	9.91	90.09	1.04
10.	Sukra	1.49	1.35	29.55	70.45	1.11
11.	Kulagara	1.29	1.23	19.79	80.21	1.05
12.	Pindrah	1.34	1.28	16.47	83.53	1.04
13.	Barar	1.61	1.50	18.30	81.70	1.08
14.	Kewai	1.40	1.33	19.06	80.94	1.06
15.	Ganta	1.21	1.19	8.92	91.08	1.02
16.	Jhurai	1.35	1.32	8.57	91.43	1.02
17.	Gauhia	1.36	1.33	8.33	91.67	1.02
18.	Kachchua	1.58	1.33	8.33	91.67	1.02
19.	Dhora	1.81	1.77	4.94	95.06	1.02
20.	Dingchi	1.15	1.10	33.33	66.67	1.05
21.	Jaiwanti	1.30	1.24	21.07	78.93	1.05
22.	Ganta (Main)	1.66	1.59	11.01	88.99	1.05
23.	Ohan	1.67	1.63	6.84	93.16	1.03
24.	Geduwa	1.38	1.35	7.81	92.19	1.02
25.	Paisuni	2.14	2.03	9.89	90.11	1.05

On the contrary standard sinuosity index and corresponding components also reveal that all the drainage lines are sinuous in nature (Table 5.4). The high percentage of topographic sinuosity index (between 66.67% and 95.06%) and the low percentage of hydraulic sinuosity index (between 4.94% and 33.33%) clearly indicate that the surface flow is principally governed with topographic factors of the region. It is observed in the field study that the surface flow paths are generally straight in manner over the hard and compact geological structure of Kaimur sandstone layered at plateau region while the course of water bodies are sinuous over the alluvium controlled plain. On the basis of the above analysis it may be stated that surface flow velocity and direction are not regular in nature. Sub-surface flow is found dominant in soluble limestone rocks. Various caves as noticed along the scarp zone validate the above face. An extensive survey has been made for the peculiar caves of Gupt-Godavari hill wherein Cave-I) and Cave-II show interesting examples of sub-surface flow solution and mechanical collapse. A continuous flow of water can be seen in Cave-II where Gupt-Godavari stream exists due to spring condition. Both the caves indicate the two different nature of dolomite structure and thus the processes involve therein are of same but acting response is in different ways. It is remarkable that the Vindhyan group is layered by limestone rocks capping with Kaimur sandstone throughout the scarp range in the region and the solution action by sub-surface water flow is most dominating in this tract. The area of Khoh Pahari about 8 km away from Karwi along the Banda-Allahabad road is affected by solution action and mechanical collapse of sub-surface flow because the area generally contains limestone rocks exposures.

The other examples of sub-surface flow processes like rilling, piping, tunnelling and gullying can be easily seen in the ravinous belt of the region. Thus, it may be stated that the vegetal cover, soil characteristics, nature of slope, topography complexity of precipitation and run-off behavior determine the sub-surface flow conditions. Most of the unvegetated barren lands characterised by sand, morrum clay and silt with occasionally blocks of sandstones and limestone structure accelerate the sub-surface flow impact.

Sheet Wash and Sheet Flow

Sheet wash and sheet flow is the process initiated by accelerated soil charged water to continuous surface flow of water swills the land surface and causes the finer soil particles to be washed away. This swilling is called sheet wash (Rabinson, H. 1972). Continued sheet wash will lead to the removal of a thin upper layer of soil which is known as sheet erosion.

A continuous thin layer of overland flow is called sheet flow. It washes mainly the particles of fine silt and clay down the slope. In the present study, it is found that sheet wash is dominating in the lower sections of the slope, rolling and smooth surfaces of northern alluvium plain along the valley sides and most of the parts of concavity formed along the isolated hills. The effective surface wash is measured under the catchment area of river Paisuni. The middle course of Paisuni denotes high intensity of sheet flow.

Rillwash and Gullying

Rillwash and gullying is a dominant fluvial process. The detailed field analysis has been prepared on rilling and gullying processes of the region. It is discussed in Chapter 9. It is found in survey that the rectilinear slopes of upland country and the valley side slopes of northern alluvium plain are badly wounded by rilling and gullying processes. The initiation of rills can be traced in most of the localities. Sinuous rills are found on the barren land having sloping surfaces and thick blanketing of sand morrum and clay deposits.

CONCLUSION

The assemblage of landforms and the terrain of Chitrakut upland is governed by a complex suite of fluvial processes. The relative importance of these processes is largely a function of climate. Ravination appears to be the most dominating process along the hillsides and river courses. Surface and sub surface processes are also dominant over the weaker sections of rock forming the upland country. Sub-surface solution is peculiar in limestone structure. Mass movement is less dominant but important on steep slopes in all over the scarp. These processes are related, directly or indirectly, to topographic factors. Slope gradient is probably a major controlling factor and on most slopes the processes vary systematically with position. Thus, climate, topography, slope and lithological characteristics appear to be the dominant controlling factors of fluvial processes.

6

Weathering and the Evolution of Landforms

In the fundamental concepts of geomorphology, it is well postulated that the geomorphic processes operating at differential rates leave their distinctive imprint upon landforms. It is also a conclusive remarks of geomorphic concepts that each geomorphic process developed its own characteristics assemblage of landforms (Prasad, G. 1987) with a full appreciation of the manifold influences of the geologic and climatic changes during the passage of time. If it is geomorphic truth that the different erosional agents act upon the earth surface, produce an orderly sequence of landforms, it may be undoubtedly said that weathering process produces various suit of landforms. Thus, the nature of weathering and its associated landforms are now the fundamental question of present discussion.

LANDFORMS DUE TO CONSTANT VOLUME WEATHERING

As mineral alteration results in the formation of new minerals of lower density than the originals, it is usually assumed that the mineral expands on weathering. Rocks being composed of minerals are assumed to be expand likewise when weathered. Commonly when a rock of the earth's surface it does expand, the change in volume causing distinctive features. However, a great deal of weathering occurs with volume change. Constant volume weathering takes place at some depth below the earth's surface (Ollier, C.D. 1969).

Ollier believes that the structures of the original rocks are found perfectly preserved despite extensive chemical and mineral alteration in most of the areas of extensive deep weathering. He reported some examples of small quartz veins where joints, small faults, original bedding, and other structural features are seen in original form in the weathered materials. Thus, in this context, it is possible to find out the evolution of landforms without volume change. On the basis of the field measurement of the region and with the help of the Ollier's experiments, the following landforms produced due to constant volume weathering can be recognised.

Constant Regolith and Saprolite

It is observed by deep weathering results that rock may be weathered to great depths. Rotten rock in place can be termed saprolite. The regolith covers both residual and transported loose or soft material overlying solid bed rock. It is found that the profile contains several zones below the ground surface but the materials of regolith and saprolite have not changed their volume but weathered minutely due to pressure, temperature and side impact. This type of example can be seen near the side exposures of rocks of Sheorampur hills. The profile of about 40 m in depth exposes residual debris of structureless sandy clay, rounded, angular, prismatic and interlocked corestones with loose morrum deposits. These corestones are seen in their original volume but small cracking without gap, colouring and decomposition in upper mineral grains can be minutely observed. This type of examples is also substantiated by Ground Water Survey Department during the course of new drilling operations in the alluvium and sandy structures of the northern plain region (Figs. 1.13 and 1.14). The geological lithologs of the region also indicate that the granite parent material which exists below 50 m in depth is more or less unchanged with few weathering signs.

Spheroidally Weathered Blocks

Spheroidally weathering generally causes expansion of rocks. The spheroidally weathered blocks are found by chemical migration of elements within the rock. Hydrolysis is the chief weathering agent of spheroidal shape of rock structure. Spheroidal shape occurs due to most rapid alteration at edges and corners of blocks which in turn with advancing exfoliation causes the original angular joint block to a rounder and spheroidal shape.

Spheroidal weathering landforms appear in the following conditions:

1. Small boulders at the surface may be weathered from all the sides by floking process and simulate spheroidal weathering. In this case the unconfined boulders would probably expand. The granite boulders as indicated near Kamad Giri Parvat, (Pahari hills near Pahari block headquarter) and Nilgarh Pahar clearly show the spheroidal shape of the boulders at ground surface. It has been minutely checked that these boulders at hill side slopes are exposed and separated with each other at ground surface due to continuous lowering of the associated platforms by erosional agents.
2. In the second case spheroidally weathered boulders may be surveyed at the same depth below the ground surface. Due to subsequent erosion, the spheroidal rock surface may be exhumed at the ground surface. Ollier suggests that these boulders do not constitute proof of formation of spheroidally weathered boulders at the surface. Once exposed at the surface, their course of weathering is likely to change to a volume-increase alteration.

These types of spheroidally weathered boulders may be traced along most of the residual hills of the upland. A peculiar example is seen near Pahari hills (Pahari block headquarters) wherein the lower profile of the slope exposes spheroidal shaped boulders. The upper section of the boulders is generally exposed at the ground surface.

Colour Banding

Colour banding appears in porous rocks (sandstone, limestone, mudstone) due to alternating enrichment and depletion mostly of iron oxides. Periodic precipitation mostly affects this landform. Ollier has observed that the banding may result from the patchy and periodic drying up of ground water with precipitation at the surface of diffuse water bodies.

The colour bands may be formed in different shape and size either following the edges of joint blocks or irregularly in various directions. They may be like chit and chips, circular, elongated, prismatic shape etc. The bands vary in thickness according to nature of porosity of rocks and the intensity and impact of periodic precipitation.

The examples of colour banding are well visualized in the limestone structure of Gupt-Godavari caves. The investigator denotes that these are formed due to continuous ground water precipitation along the edges of the limestone blocks. The bands are generally marked along the edges of the rocks in the form of chit and chips.

Joint Hardening and Softening

Joint hardening and softening depend upon the nature of the rocks and its joints and fissures characteristics. Water moves with dissolved ions along the joints and fissures of the rocks which, in turn, initiate weathering processes. If the impregnated joints are harder than the blocks, later erosion may give rise to a boxwork type of feature. This type of joint hardening and softening is visualized in the western part of the Kamad Giri Parvat.

EXPANSION WEATHERING AND CAUSATIVE LANDFORMS

Expansion of rocks causes due to temperature changes. Thermal expansion and contraction of rocks cause disintegration. Dark minerals absorb heat faster which may also give rise to differential expansion leading to many small stresses within a rock. Differential expansion may lead to the formation of minute cracks and possibly granular disintegration. The landforms of expansion weathering form in the following conditions.

Landforms due to Unloading

Unloading is the process of massive exfoliation (dilatation) that gives rise to slabs of rock several feet in thickness and associated landforms (Ollier, C.D. 1969). These landforms are mostly recognised in granites and partially seen in other igneous rocks, metamorphic and in some sedimentary rocks. Unloading may cause the following landforms.

Unloading Sheets

Unloading process initiates to produce a large and thick curved slabs of rocks which contains an exposure of the dome-shaped rock beneath. These sheets may be termed as beds of joints. It is also known as topographic jointing. These partings are mostly found according to sedimentation and parallel to the ground surface.

The plateau region illustrates several examples of unloading sheets. The Paisuni river side scarp which is well jointed by Kaimur sandstone and Tirohan limestone rocks is the best example of topographical jointing. Sati Anusuiya Gorge, and Limestone jointing near Gupt-Godavari hill generally depict the peculiar view of unloading sheets in sandstone and limestone structure.

Unloading Domes

Unloading domes is the name given to those convex hills of hard rock capping which are usually devoid of vegetation and partly covered with broken unloaded sheets. The hillock country of Yamuna plain shows such type of domes. These domes are vegetationless and formed on older granite rocks. The upper free face is washed by various agents of weathering and erosion. These domes occasionally contain granite boulders and scattered loose materials of topographical sheeting. The northern section of Pahari hills near Pahari block proves the existing domes condition.

DIFFERENTIAL WEATHERING AND ASSOCIATED LANDFORMS

Features due to differential weathering are numerous. Differential weathering occurs by structural or lithological differences in rock. Irregular weathering processes also initiate landforms differentiation on homogeneous rock characteristics. Random patches of vegetal properties on limestone, granite or some other lithological units may also give rise the assemblage of landforms due to differential weathering. Thus, it may be said that differential weathering can occur on all sides. At the upper end of the scale differential weathering and erosion make birth of valleys and hills of multiple character. It registers various micro-level regional landforms which are not very easy to recognise in the field measurement. It is also remarkable that both micro and macro weathering landforms have their equal significance. Micro features may be several in number on one macro landform. Thus, the small features are the cause and effect of the larger sized landforms. Several geomorphologists have suggested some example of micro-features like ground level platforms, boulder streams, weathering pits, boxwork, honeycombed structure, hoodeo, butcher block, raised rim, ball and buns gobular mass of differential weathering etc. produced by differential weathering. The measurement of micro-features is difficult task and not presented in the text. The examples as quoted by above photoplates have been assessed on the basis of foreign literatures and the photographs given as the proof.

Landforms Due to Sculping

Sculpted Granite Blocks

Sculping refers to the surface differential weathering only, not subsurface alternation. Many corestones become sculpted when they are exposed to surface weathering and erosion. It causes an individual boulders into a variety of shapes that almost defy classification. The shape, spacing, orientation of depressions and rises seem to be of endless variety (Ollier, C.D. 1969). It occurs in granite rocks and the intensity of weathering appears quite different in the same area. A peculiar example of sculping in harder granite rocks can be seen in the larger blocks existing in the western flank of Kamta Nath hill where blocks are of different size and shape. But the softer granite in the same area usually shows the different characteristics of weathering due to sculping impact. It may be said that the sculping may be caused different characteristics of rock features.

LANDFORMS DUE TO STRIPPING

Due to stripping, blocks of unweathered rock may float in the regolith as corestones. The following landforms are thought to be formed by stripping.

Tors and Related Landforms

Following to Ollier, C.D. (1969) Tors are small hills or peaks of boulders usually about 20 to 60 feet high rising abruptly from the surrounding gentle ground surface. Tors appears during the course of erosion when the basal surface of weathering may be exhumed and the corestones left behind as the soft material of the regolith is washed away. Stripping and subaerial weathering cause the formation of tors and associated landforms. It may be formed by one cycle but many are formed by two cycle process of deep weathering followed by exhumation. Tors are most common as an isolated exposure of much jointed granite rock, standing as a prominent castellated mass above the general surface of a plateau.

Tors may be formed in different climate at different height. The size and shape of the boulders may vary according to rock structures and tors forming processes.

The detailed study of tors may be considered in Chapter 8. It is notable that the tors are found mostly in hard granite rocks of the area. A view of Bharatkup hills, Kamta Nath hills, Pahari block hill, and Bharatpur hills exhibit the peculiar examples of Tors in the region.

Stripping Sinkholes and Caves

Sinkholes are the depression generally caused by underground water solution action over the limestone rocks. But stripping gives rise some peculiar sinkholes which in turn cause cavity along the joints. Ollier, C.D. (1965) has introduced a granite cave from Victoria, Australia which is formed by stripping processes.

A typical example of sinkhole and cave is observed in sandstone country about 1 km north-west of Jamunhai village located along the scarpline. On topographical sheet, it is marked as Birudh Kund. It has a diameter of 75 to 90 m and a depth of 75 m in sandstone rocks. Hukku, B.M. (1966) reported that the sinkhole is formed by the collapse of the weathered rocks. Now the kund is 150 m in depth as noted by recent topographical sheet. The investigator opines that the formation of Birudh kind is concerned with stripping sinkhole and finally due to collapse of rocks, it is converted into a big size kund or open cave like Uvala.

LIMESTONE WEATHERING AND SOLUTIONAL LANDFORMS

Thornbury, W.D. (1954) has proposed a useful classification of karst landscapes. He denotes major karst regions in which all the landforms have been affected by solution as the dominant agent creating landforms. Minor karst regions are those in which solution features are present but do not dominate the landscape. The present study region falls under minor karst region and thus the karst landforms are notable but not dominating rather than others.

In limestone area (karst region) solution gives rise to many landforms. The terminology applies to these landforms is vast and complicated because the different karst region evolves its own landform names and many words from many languages have been used as technical terms.

A brief sketch of solutional landforms in limestone area of the region is depicted as follows (Prasad, G. 1986).

Terra Rossa

Terra rossa is a team applied loosely to any fine textured and fairly plastic clay that acquires a natural vitreous skin in burning and that is used in the manufacture of terra-cotta. It contains a peculiar brownish-red or yellowish-red colour. The descending ground water usually leaves a residue of a red clayey soil mantling the surface and extending down into open joints in limestone area by surface and near surface solution. The loose materials are known as terra-rossa. It may be found on the rock surface of moderate and gentle slope in varying thickness. Plates 34, 94 and 95 indicate yellowish-red colour of terra-rossa resting surrounding the limestone rocks of Khoh Pahari located about 8 km away from Karwi along Karwi-Allahabad road.

Sinkholes and Associated Forms

Sinkholes are the most common and wide spread topographic forms of limestone area due to solution. It is impossible to check the intensity of sinkholes in the limestone area. It depends upon the nature of limestone rocks and the intensity of solution and weathering. Swallow holes, polje, doline and uvala are the other features developed by sinkholes. In the present study region, the other associated landforms are not seen. A compound sinkholes are seen over the top of the Gupt-Godavari hill (about 280 m in height). These sinkholes are found in a series between the course of a fingertip tributary. The tributary looses their water near the sinkhole surface and thus the platform defines the formation of 'sinking creeks'.

Cave

Cave formation is a typical topography in limestone area. The formation of cave depends upon the solution of ground water and thereby caused weathering impact on existing limestone. The nature and characteristics of limestone rocks are also the dominant factors of cave formation. In the present study region, the carbonate rocks which are most common along the scarp zone from Gupt-Godavari to Khoh Pahari (east of Karwi) give rise to different peculiar landforms. Solution caves are found near Gupt-Godavari in the south-west of Chaubepur village, Madhwa hill (south of Brijlal Ka Purwa), and around Bankesidh (2 km south of Sidhpur village towards south-east of Karwi). The caves of Madhwa hills and Bankesidh are not so important as the Gupt-Godavari caves. The detailed study of cave formation may be discussed in Chapter 8.

These caves incorporate various micro-level landforms due to carbonation processes. The recognised features are discussed here.

Cave Pillar

Cave pillar generally forms due to joining of downward extending stalactites and the upward growing stalagmites in the cave area. But pillars may also be existed in hard and compact dolomitic area due to solution of ground water. A peculiar example of cave pillar is seen near Sriram Kund area where soluble carbonate rocks have been washed away from all sides due to spring condition and a hard limestone rock has been appeared as a cave pillar. Cave pillar is seen in cave II of Gupt-Godavari hills.

Stalactites and Stalagmites

Stalagmites are the pendant mass of calcite hanging vertically from the roof of a cave. It is a cylindrical or conical deposit of mineral matter sedimented from drops of water containing calcium bicarbonate which have seeped through crevices and joints of limestone rocks. The roof of the cave I and cave II of Gupt-Godavari hills indicate several examples of the smaller stalactites. Stalagmites are the similar mass to a stalactite growing upward from the floor of cave and below the stalactites by dripping water. It is also composed of calcite. In Gupt-Godavari cave II, there are no signs of stalagmites due to flowing water but in cave I only loose materials of calcite are noted in place of stalagmites on cave floor.

It is remarkable that both the caves indicate well developed 'DRIP STONE' along the cave roof in spite of stalactites. The cave roof and walls (cave I) are also marked with columns. 'FLOW STONE' and 'RIM STONE' are common features in cave II where stones move with flowing water in different size and shape.

Helictite

Davis, W.M. (1930) suggested another feature in limestone cave namely as helictite which is most common to stalactite. Helictite may be formed by deposition of calcite from solutions percolating down fine fissures on to ceilings of cave in limestone country usually where a strong current of air is present. It may be thin as a thread, in spirals or loops or festooned with tenacles. It is believed that the growth of helictite does not necessarily extend along vertical lines. Its individual parts may grow upward, horizontally, obliquely or in curves as well as downward. It is not certain why helictites are apparently able to defy gravity but the most logical explanation seems to be that they develop where water is not entering the cave in sufficient quantity to give rise to drops that fall. If there is only enough water to keep the surface wet, then growth of the individual parts of the helictite depends upon chance orientation of the crystal axes of calcium carbonate. This may be in any direction.

The present investigator believes that most of the stalactites as discussed earlier are helictites because very few of them contain water. The hanging calcites of cave II are more probably helictite.

Blind Valley

Blind valley is a typical feature of karst region wherein valley ends abruptly downstream at the point (stream base) where the flowing river disappears underground. This is caused either by the collapse of the roof of an underground stream course or the grading of a surface stream to a progressively falling base-level. This type of valleys are surveyed over the top of the Gupt-Godavari hill where the compound sinkholes are existed and in the lower section of Gupt-Godavari hill where dry bed of the river is observed due to sinking of the flowing water of Gupt-Godavari cave.

CONCLUSION

On the basis of the above discussion, it may be concluded that the weathering landforms are principally governed with mechanical (the disintegration of rock without change in chemical composition and chemical (the disintegration of rock with change in chemical composition) actions. The landforms caused by mechanical weathering are associated with insolation, forest action and by the prying action of plant roots.

The chemical weathering landforms are chiefly concerned with surface and subsurface water solution. Water is a potent destructive agent. It causes the disintegration of rocks by chemical reactions. Oxygen, carbondioxide and organic acids are the chemical constituents that support chemical reactions in rocks. Few landforms also exist by mineral composition in rocks. Thus, it may be stated that the landforms are the result of combined attack of chemical and mechanical processes wherein chemical impacts are predominating. As a matter of fact, weathering by itself cannot produce major landforms.

In real sense, landforms came into being only when the weathered material leaves its place of origin. The material movement occurs through the action of denudation and erosion which are motivated by gravity. This is also true that weathered materials move by different erosional agents following the shortest possible route in the direction of gradient. These erosional agents (specially erosion and mass movement) remove the solid particles loosened from the rock mass as soon as these come into the paths of movement by weathering processes.

Thus, it may be considered that the landforms as discussed in this chapter are not produced by single process whether it is weathering or erosion. The fact is that weathering prepares the way for the processes, erosion and mass movement to operate in producing landforms. Landforms are the result of structure process and stage. One dominating factor associating with many others cause the landforms of significant structure with the passage of time.

7

Fluvial Topography

The present chapter incorporates the study of fluvial topography wherein the general cyclic stages and presentday topography, fluvial processes constituting slopes and scarps, erosional features, nature of valley development including long and cross valley profiles, development of waterfalls and springs and the formational characteristics of depositional landforms have been discussed systematically. The general outlook of regional topographical complexities have been assessed in view of geological history of the region whereas the cyclic stages of the region have been considered with previous comments and constituting the streamline surfaces with the help of drainage and contour maps of the region.

GENERAL CYCLIC STAGES AND THE PRESENT TOPOGRAPHY

The Chitrakut and its adjoining areas as an essential part of Vindhyan ranges form a conspicuous landscape. The geological structure and lithology have played an important role in giving rise to the geomorphic forms of Chitrakut. The sandstones mostly form ridges and plateaus in the south whereas the alluvium deposits maintain gently rolling plain of Yamuna-Paisuni river in the north. The Chitrakut plateau rises abruptly from the plain stretching in an approximate ENE-WSW direction. A few small peaks of this plateau like Sita Rasoi (303 m), Jatari Baba Thar (343 m) and Gupt-Godawari (307 m) form prominent landmarks. To the north of this scarp, the adjoining topography is undulatory dotted with numerous knoks like Kamtanath (25° 10′, 80° 51′), Sangrampur (25° 12′, 80° 50′), Sheorampur (25° 10′, 80° 46′), Bharatpur (25° 12′, 80° 48′) and

Lodhwara (25° 15′, 80° 55′) (Kedar Narain 1960). Towards the north of the scarpline, hillocks of the basement granites, occasionally capped by the Vindhyan rocks and surrounded by sub-recent to recent alluvial sediments, are found. The scarpline is well dissected by denudational processes having limited vertical cliffing overlying rectilinear and concave slope profile in plan. In the extreme, south, the plateau region shows gorge position along the upper reaches of river Paisuni. In the west of the study region, the plateau regime is well dissected with the result a peculiar example of dissected hills are noted.

The Kaimur sandstone forms the hilltops, underlying which a regular band of Tirohan limestone is marked. Tirohan limestone is cavernous in nature and thus the formation of limestone cave has became possible in the region. Seepage formation and spring conditions are also marked along the sandstone limestone structure throughout the scarp zone. Waterfalls and rapids are also marked along the riverbed of Paisuni mostly in their upper reaches due to structural control. The rectilinear slope profile of the scarpzone generally in the eastward and westward of Gupt-Godavari hills is ravaged by rills and ravines. Pilling and ravination are also most common along the foothill zones riverbed sides and over the most of the barren sloping ground of dissected hills located in the north of the scarpzone. Thus, the entire view of Chitrakut topography reflects senile characteristics of landscape development where the denudational processes have highly consumed the initial landscape specially through slope declining, parallel retreating and backwearing of the scarps.

Roy, A.K. and Bhattacharya, A. (1982) have stated that the present day positive topography of the Vindhyan was the site of huge ancient inland sea. Due to gradual uplift and change in palaeogeomorphic set up, the present topography has evolved. The palaeogeomorphic set up was that of a barrier coastline with morpho-units like barrier beach dune, tidal flat and lagoon. In the presentday topography they are represented by ridges and plateaus and intervening valleys. The uplifted Vindhyan topography was later subjected to the process of denudation, scarp retreat and peneplanation giving rise to various landforms. Quite later, the Vindhyan formations and topography were covered under profuse outpourings of the Deccan laves which had indirect influences in this region also. Denudation has exhumed the Vindhyan topography which continues backwearing of scarps even to the present day.

Roy and Bhattacharya further advocated that the Vindhyan sedimentation was uplifted and peneplaned in several phases between late pre-Cambrian to early Cambrian and even after. The evidence of peneplanation is afforded by the presence of many flat topped hills at different levels having the same elevation. These flat topped hills are dissected and detached from one another forming pediment in between. The remarkable geomorphic evolution of the palaeo-landforms of the fluvio-marine origin into the present day landforms comprising plateaus and Cuesta-hogback ridges took place as a result of gradual uplift of the basin. Because of the gentle synclinal structure, well developed Cuesta landforms were formed discernible throughout the Vindhyan landscape.

Bhatnagar, N.C. and Sah, D.L. (1966-67) have stated that the geomorphological features of the areas north of the Vindhyan scarp seem to suggest high relief bed-rock topography which may lie submerged under a cover of sub-recent, sediments. It may also contain ancient buried channels filled with previous materials. Based on the results of seismic soundings by the Geophysical Department of U.P. Govt. the possibilities of such burried bed rock channel fills beneath the clays have been indicated. The continuity of such channel-fills towards the north may be traced by extensive geophysical surveys on a regional basis.

The structural formation clearly indicates the cyclic stages of the region. The Bundelkhand pediplain represents an old erosion surface carved out of granite. Three planer surfaces with an average elevation of 100 m, 150 m and 300 m characterize almost the entire province (Roy, A.K. and Bhattacharya, A. 1982). The Tirohan formation overlying Bundelkhand granite indicates the second cyclic stages of landform development which is well proved by the unconformity located between Tirohan limestone and Kaimur sandstone. The Kaimur and Rewa group separately confirm the later cyclic stages of landform development. The alluvium cover of north plain advocates the recent platform. These geological formations, thus represent the different cyclic stages of landform development.

In view of studying landforms and erosional surfaces, Dubey, R.S. (1968) has suggested three erosion surfaces developed in the plateau region namely as (i) the post lower Vindhyan erosion surface developed at the top of the lower Vindhyans; (ii) the post-Kaimur surface developed at the top of the Kaimur series; and (iii) the post-Rewa-erosion surface developed at the top of Rewa series. These three erosion surfaces are exhumed and can be determined by unconformities resting between the two distinct formations.

Taking into consideration of Altimetric frequency histogram curve and Area height curve of the region (Fig. 2.12) the following erosion surfaces can be recognised:

1. *420 m Surface:* The top surfaces of Bijehna and Gerua pahar located in the southern part indicate 420 m surface.
2. *360 m Surface:* Upper reaches of Ghinhal basin, top surfaces of Kararis pahar and Morpho fort exhibit this surface.
3. *300 m Surface:* North facing scarp and hill tops between Paisuni and Ohan catchment area represent the above surface.
4. *200 m Surface:* The hill ranges in the east of Ohan river show 200m surfaces. These are remnants of former erosion surface.
5. *100 m Surface:* The undulatory flat surface of northern alluvial plain suggest 100 m surface. It is a depositional surface constructed due to erosional agents of the region.

At attempt has been made to prepare the stream line surfaces (Fig. 7.1) with the help of drainage lines and contour maps of the region. Streamline surfaces of 100 m, 180 m, 260 m, 340 m and 420 m indicate various plantation surfaces and stages of cycle of erosion in the region.

FLUVIAL PROCESSES AND DEVELOPMENT OF SLOPES AND SCARPS

The study region is fluvially dominated where fluvial processes have actively consumed the land and surface irregularities have initiated at a greater extent. There are 2838 drainage lines wherein 2154 are of first order, 543 of second order, 116 of third order, 23 of fourth order and 03 are of fifth order. The Paisuni river represents six order. Thus, the total surface area of 1851.402 km^2 registers 1.533 drainage lines at per km^2 which is also the indicative of the dominance of fluvial processes.

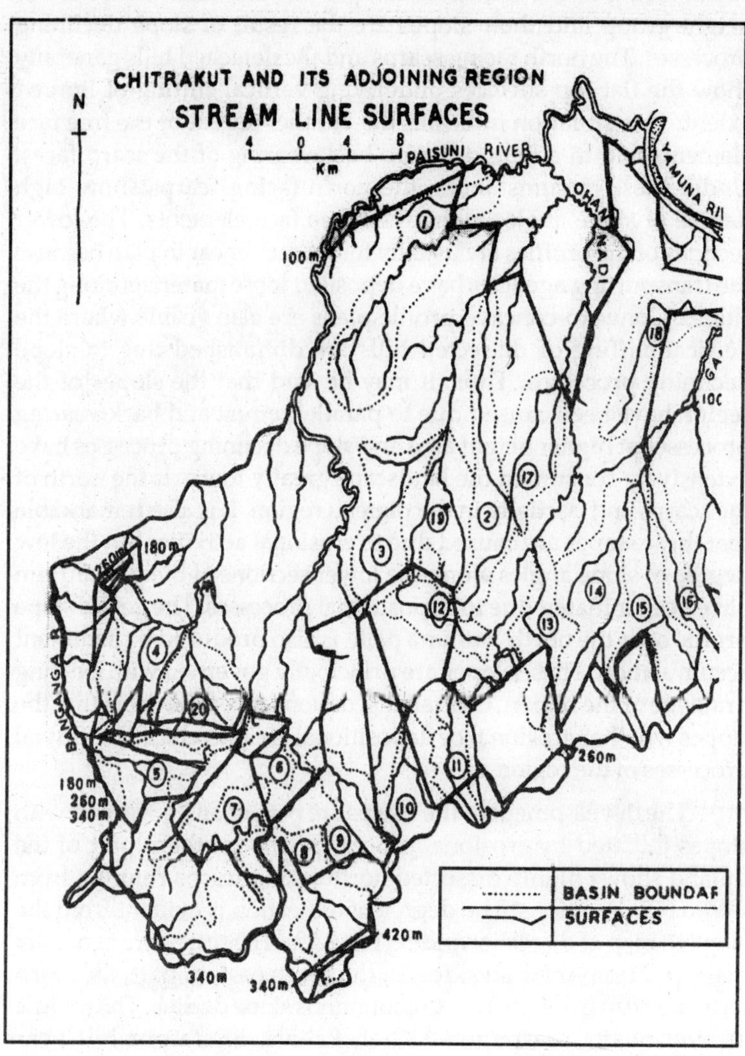
CHITRAKUT AND ITS ADJOINING REGION
STREAM LINE SURFACES
N
4 0 4 8
Km
PAISUNI RIVER
YAMUNA RI
OHAN NADI
100m
BANGANGA R.
180m
260m
180m
260m
340m
260m
420m
340m
340m
BASIN BOUNDAF
SURFACES

Fig. 7.1

The presentday topography as represented by ridges, plateaus, and scarps are the results of upliftment, structural deformation and denudation through ages. The presentday slopes and scarps have also generated through parallel retreat and backwearing of the initial landscape. The residual hills of lower height group and their slopes are the result of slope declining processes. The north facing scarps and the detached hills generally show the flat top surfaces underlying vertical cliffing of limited extent. This situation indicates the parallel retreat of the free face elements and in a long duration backwearing of the scarp faces. Under these circumstances, the north facing scarps show high degree of slope angles along their free face elements. The lower sections of the profiles are usually found rectilinear in plan because the transporting agencies have deposited loose materials along the slopes. Convexo-concave profile plans are also visible where the vertical cliffing of dissected hills are diminished due to slope declining processes. Thus, it may be said that the slopes of the region have been caused due to parallel retreat and backwearing processes at higher height area and slope declining processes have extensively consumed the landscape greatly towards the north of the scarps and partially in the plateau region. It is also remarkable that these slopes are caused due to erosional activities but the low degree of slope angles along the lower sections of the plateau rim have been initiated due to depositional processes. The gentle slope profile over the north Yamuna plain is also produced by sediment accumulation. These slopes are principally governed with existing drainage of the region. Overall discussions reveal the facts that the slopes whether erosional or depositional are introduced by fluvial processes of the region.

The development of the scarps are principally governed with slopes initiated by erosional processes. The southern part of the upland shows highly dissected north facing scarps running from SW to NE directions. The degree of dissection is evident from the isopleth map of dissection index (Fig. 2.22). The scarps are generally dissected from their all sides. In the south eastern part, the scarp faces are 60 m in height due to continuous slope decline. The middle section of the scarps near Khoh Pahari, Bharatpur hill near Bharatkup, surrounding area of Bharatkup, Sheorampur a view of Hanumandhara scarp near Sitapur and a view of scarp in the north of Sheorampur are also much dissected and of about 160 m in height. The lower part of these scarps is either rectilinear or concave in plan and generally shaded by vegetal cover. The upper parts denote sandstone capping initiating cliffing and free face section overlying flattops.

In the south and south-western part, the scarps are characterised with sandstone and limestone formations. The northern phases are deeply dissected and along the Paisuni river course they are defined by narrow cut deep gorges. The vertical wall of the scarps is most probably caused by vertical erosion and structural stability of the mass. The Sati Anusuiya scarp shows free face element with massive sandstone. The scarps near Gupt-Godawari and in the south-western part are also found massive in geological formation and about 160 m in height above ground surface.

The field survey reveals the fact that the steep scarps prove the parallel retreat of slopes without sufficient loss in gradient. Even the flat rolling surface between the detached hills and the plateau rims may not be taken as a result of lateral plantation rather they are the outcome of backwearing of hill sides due to parallel retreat of slopes. The presentday scarps have developed a well defined convexo-concave slope profile near Hanumandhara hills. Free face, rectilinear and concave slope profiles are also measured.

The most outstanding feature of the scarps is that they are under effect of continuous process of parallel retreat generally in the south and south-western part maintaining their maximum slope angles. The recession of the scarps has taken place and is still active under the impact of weathering of the scarp faces and constant removal of debris. Various detached hills exhibiting the example of Mesa, Buttee etc. projecting above the general rolling uplands are the left over remnants of the recession of the scarps. The area (in the northern part of the Sheorampur) and (Lakshman Pahari near Kamadgiri hill) are the examples showing the detached hills. In the north of the scarp faces the escarpments along the detached hills cover loose materials having moderate slope and minimum height.

Thus, the slope evolution of the presentday topography can be explained by composite theory of parallel retreat, slope replacement decline. Single process cannot be formed the complexity of slopes of the region.

OTHER EROSIONAL FEATURES

Mesas and Buttes

Mesa is a flat-topped eminence (table land) existing as the remnant of denudation of a plateau and commonly copped with a resistant rock stratum. Butte is a small isolated portion of mesa. These landforms exist between the assemblage of detached dissected hills. Most of the area away from the scarplines register several peculiar examples of detached hills and defines the nature of mesas and buttes.

The panoramic view of Kamtanath hill (300 m in height and conical and circular in shape), Hihara hill (260 m in height and first topped), Nilgarh Pahar (312 m in height elongated with flat topped), Nilgarh Pahar (312 m is height elongated with flat topped, Baunri hill (258 m in height conical in shape), Itwan hill (231 m in height with flat topped), Morpho Pahar (379 m), Hadiyar Pahar (310 m, elongated in shape), Barra and Garra Pahar (295 m), Chandra Maheshwari (332 m like Cuesta) and Gonda ka pahar (306 m, like Cuesta) are the best examples of mesas. Ghosh, D.K. (1978) has stated that the isolated hills found in the country are sharp-edged cliff faced and termed as inselberg of almost accordant heights and are surrounded by gently sloping plane in bedrock called pediment.

It is observed that these mesas and buttes are of very sharp edges having steep slope in the south-western part while in the south-eastern flank (near Karwi) these are visualized in moderate slope plan. Lakshman Pahari (219m, Sangrampur hill and surroundings (245 m, 25° 11′ N lat. and 80° 50′ E long. Muradpur hill (218m, 25° 7′ N lat. and 80° 41′ E long.) and Kari Pahar (212 m, 25° 12.5′ N lat. and 80° 43′ E long.) are the typical examples of Buttes and have very limited flat topped surfaces characterized by pronounced free face elements.

Water Divides

Water divides are the source regions (birth place) of tributaries flowing accordant to slope. It is determined by the highest elevation of contours rising at the top of hill ranges. It becomes higher than the surrounding plane. The catchment area of the basins are demarcated by water divides.

Two major water divides (Fig. 1.15) are recognised in the region for example: (a) running from south-west to north-east direction along the Allahabad-Satna central railway lines in the south-east of the study region where Jaiwanti, Ganta, Kewai, Barar, Ohan, Maro and Kalibarah rivers take their sources and join the major drainage line flowing towards north-east (in the case of Ganta and Ohan's tributaries) and north-west (in the case of tributary streams in the right flank of Paisuni river) directions; (b) the second water divides is marked in the south-western part of the region where Banganga (a tributary of Baghain river) and other small tributaries descend towards north-west direction. These water divides make the boundary of the region. The northern scarp faces also act as a major water divides running from south-west to north-east direction. Barwa, Dingchi (left flank of Paisuni river), Piprawal, Gauhia, Paroi and Dhora (between the catchment of Ohan and Paisuni rivers) take their sources in the south and south-west and descend through northern alluvial plain divorcing the scarp zones. The source region of the aforesaid rivers are water dividing zones of the region. The top of the isolated hills may be also recognised as water divides. In Fig. 1.15, the major drainage basins are separated. The boundary of the basins are the true indicator of water divides. It is remarkable that the water divides form with the upliftment of the place surface in the initial stage of cycle of erosion.

Residual Isolated Hills, Erosional Remnants (Monadnocks) Rock Exposures, Domes etc.

Intra-basinal water divides and inter-basin hillocks of the northern plain have been wasted and worn back by lateral planation ending into several monadnocks surrounded by fertile plains. These monadnocks are varying sizes at certain places, bear exposed granitic domes and somewhere are covered with regoliths. The rolling surface near sitapur clearly indicate the formation of monadnocks surrounded by alluvial plains.

Hill systems of the study region have been lowered down and rounded by denudational processes and thus the isolated hills are generally of rounded in shape. Few examples like Chitrakut plain in the north of Hanumandhara hills area near Sitakund rounded hills near Sheorampur and isolated hills in the western part of the region located in Satna district explain the formation of isolated rounded hills of the region. Conical hills are also observed in the region. It can be well recognised near Bharatkup Lakshman Pahari near Kamadgiri Parvat, Sitpur and in the western flank of Sitapur. These hills are equally eroded from all the sides and contain harder rocks over the top surfaces. Erosional remnants are seen in the upper reaches of Bhurra nala (a left flank tributary of Kewai and located in the right flank of Ohan nadi near Ohan-Barar confluence), the right and left flank of Bhaisondoh nala (a tributary of Paisuni) and the left flank of Geduwa nala (near Itaura village) are marked by erosional remnants.

It is also remarkable that the residual isolated hills are characterized by convexo-concave-rectilinear slope elements and are scattered over the northern lowland region mostly adjoining to the plateau rim. Due to excessive fluvial erosion rock exposures tending towards north are frequently found in the region. These exposures are occasionally seen in the river beds near Raghav prayag ghat, Paisuni river. They form rapids as seen near Pramod ban. Granitic domes are seen near Pahari block.

Structural Benches

Structural benches are formed where the horizontal bedding of hard and soft rocks is found alternately. The soft rocks easily washed away by denudational processes between the harder rocks and thus the compact bedding of hard rocks makes valley benches. The structural benches differ to alluvial terraces because these are made by structural control. These benches may be observed either one or both the sides of the valley according to nature of the geological structure. Valley benches are seen near Bansachua located in Manikpur hill ranges along Manikpur-Satna railway line. It may be recognised that the benches are made by harder sandstone rocks alternatively bedded between softer rocks along the tributary streams. The softer rocks have been washed away due to continuous flow of water by spring conditions located there. Following to longitudinal profiles of major drainage basins (Fig. 7.2) of the region, it may be explained that the structural benches can be formed among the drainage lines due to structural heterogeneity. The longitudinal profiles of Kalibarah, Jhurai, Ghinhal, Maro, Pindrah, Ganta, Ohan and Paisuni rivers clearly indicate break in slope from highest elevation to valley bottom. Undoubtedly, these breaks are the sites of water falls and prove the rejuvenation of the terrain. Thus, these basins make structural benches between their source and mouth.

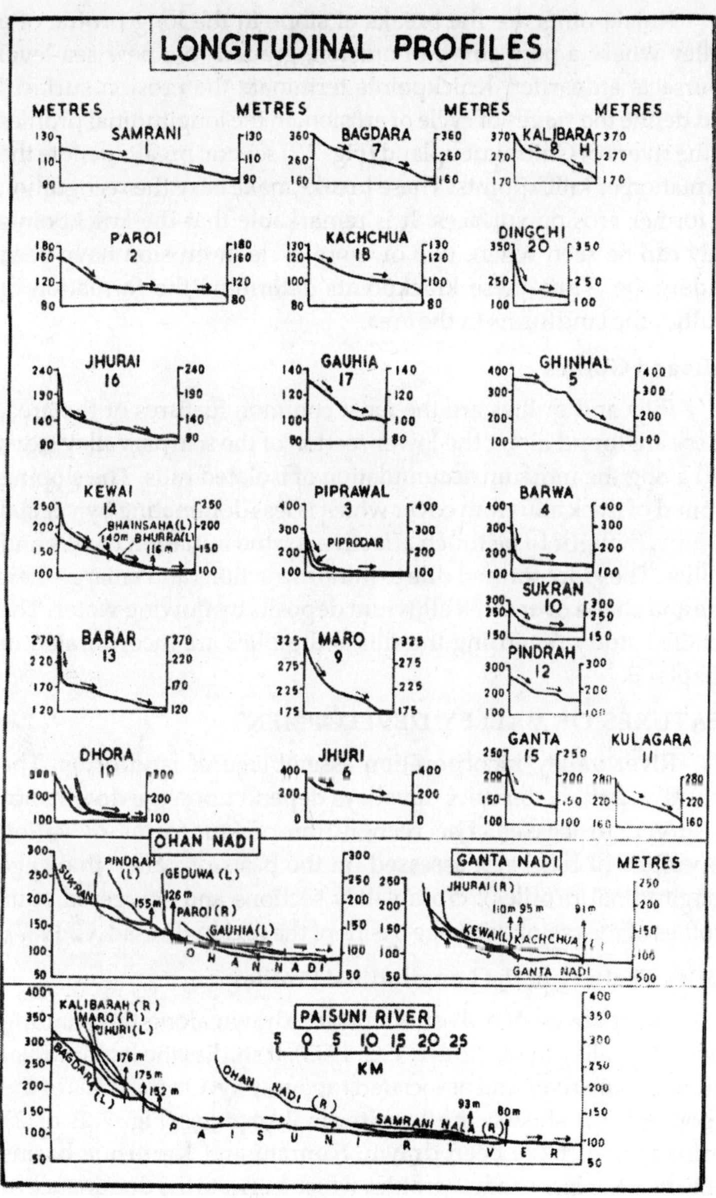
LONGITUDINAL PROFILES
METRES
SAMRANI
BAGDARA
KALIBARA
PAROI
KACHCHUA
DINGCHI
JHURAI
GAUHIA
GHINHAL
KEWAI
BHAINSAHA (L)
BHURRA (L)
PIPRAWAL
PIPRODAR
BARWA
SUKRAN
BARAR
MARO
PINDRAH
DHORA
JHURI
GANTA
KULAGARA
OHAN NADI
GEDUWA (L)
PAROI (R)
GAUHIA (L)
GANTA NADI
JHURAI (R)
KEWAI (L)
KACHCHUA
PAISUNI RIVER
KALIBARAH (R)
MARO (R)
JHURI (L)
BAGDARA (L)
OHAN NADI (R)
SAMRANI NALA (R)
KM

Fig. 7.2

Knickpoints

Knickpoints are the breaks of slope in the long profile of a valley where a new curve of erosion, graded to a new sea-level intersects an earlier. Knickpoints terminate the erosion surfaces and define the stages of cycle of erosion. In the longitudinal profiles of the rivers of Chitrakut upland (Fig. 7.2) several breaks denote the formation of knickpoints. These breaks make easy the recognition of former erosion surfaces. It is remarkable that the knickpoints only can be seen where two or more cycle or erosion have been undergone. Thus these knickpoints determine the formation of multicyclic landforms in the area.

Rills and Gullies

Rills and gullies are the most common features of the area. These are found along the lower section of the scarps, valley sides and along the morrum accumulation of isolated hills. The sloping ground of thick alluvium cover which is less dominating by vegetal cover is thought to be much affective for the initiation of rills and gullies. These are formed due to raindrop actions and enlarge their size and shape over thick alluvium deposits by flowing water. The detailed study regarding the rills and gullies are incorporated in Chapter 9.

FEATURES OF VALLEY DEVELOPMENT

River valley incorporation assemblage of landforms. The nature and characteristics of valleys depend upon the dominance of fluvial processes. The recognition of landforms of valley development has been assessed on the basis of valley thalwegs (longitudinal profiles), cross valley sections and staggered spur profiles of the major drainage basins of the region (Prasad, G. 1987).

Valley Thalweg or the Longitudinal Profiles

The thalweg of a river is a profile drawn along the winding line of the valley floor (Miller, V.C. 1953). It studies the longitudinal courses of the river and associated topography. On the basis of the topographical sheets, the longitudinal profiles (Fig. 7.2) of 23 drainage lines have been drawn. Samrani and Kachchua basins show more or less graded profiles which indicate the dominance of depositional landforms along the valley. Basins like Paroi, Dingchi,

Jhurai, Piprawal, Gauhia, Barwa and Dhora exhibit steep slopes along their upper reaches and gentler slopes along their middle and lower profile sections. This is the indicative of high relief topography in the upper sections and flat plain country and depositional landforms along the middle and lower profile sections. Some minor fluctuations of relief are marked in Barar having steeper slope in their upper part. Basins like Jhuri, Sukra, Kulagara and Bagdara show minor irregularities in relief and associated topography. The remaining valleys represent breaks in slope and relief irregularities which in turn cause landforms assemblage in the valley region. The study of profiles indicate the fact that the river basins which are developed in the flat plain area have less topographical changes while the basins partly draining over plateau and partly draining over plain preserve variety of landforms due to lithological heterogeneity and dominance of fluvial processes.

Valley Cross Sections

Valley cross sections (Fig. 7.3) illuminate the nature and characteristics of denudational processes and valley forms. It makes an evidence to recognise the stages of cycle of erosion, erosion surfaces, breaks of slopes and knickpoints. The cross valley sections of Samrani, Kachchua and Jhuri basins have not been drawn due to flattish terrain. The cross sections of Barar, Gauhia, Piprawal, Dhora, Paroi and Barwa basins clearly show the flat plain country with meandering course of the river. The upper concavity of Barwa is very much revealed on cross section. Paroi, Piprawal and Jhurai basins slightly indicate v shaped nature in their upper reaches. Ghinhal indicates a clear picture of erosion surface, knickpoints and breaks in slope. Kalibarah basin shows deep linear gorges in their lower reaches and makes the valley in v shaped at the height of 270 m. Bagdara, Sukra and Kulagara basins indicate symmetrical cross sections of the valley. Kewai basin exhibits terminal points at the height of 160 m and in the lower height area, it maintains wider valley on plain. Ganta shows a significant break at the height of 160 m. It is also remarkable that the basins draining over the plateau initiate the valley forms in v-shaped or asymmetrical in nature. But in the flat plain country these basins form symmetrical valley forms.

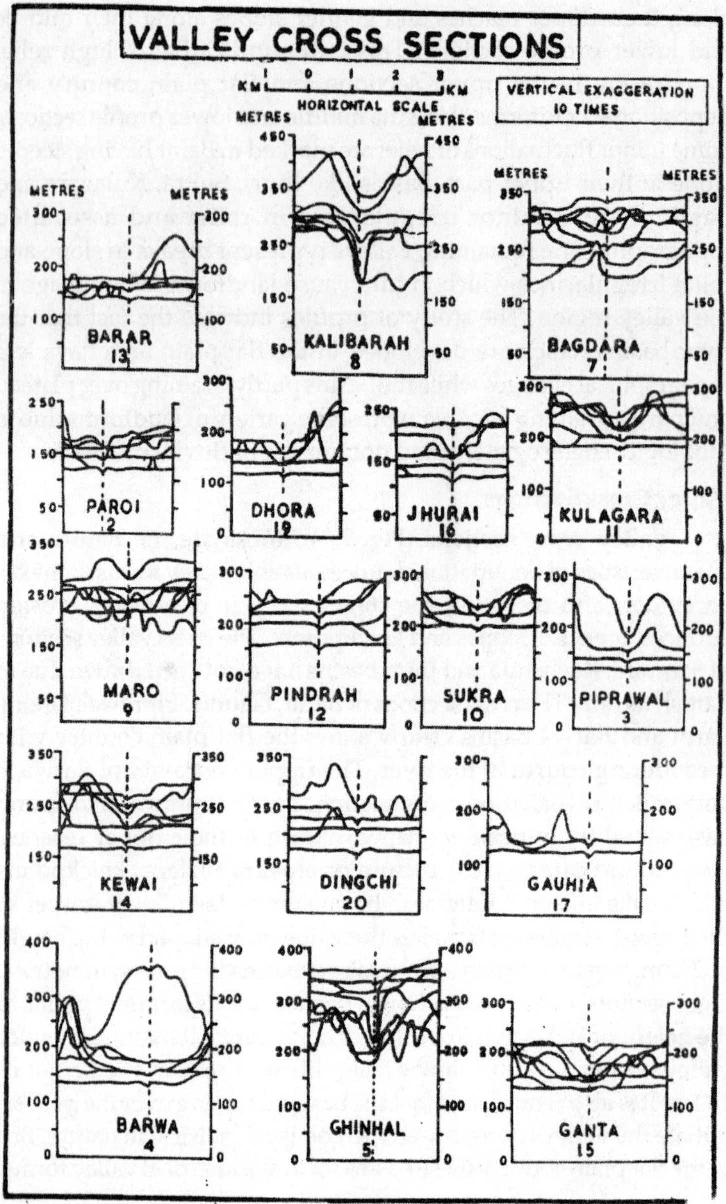
VALLEY CROSS SECTIONS
HORIZONTAL SCALE
VERTICAL EXAGGERATION 10 TIMES
BARAR 13
KALIBARAH 8
BAGDARA 7
PAROI 2
DHORA 19
JHURAI 16
KULAGARA 11
MARO 9
PINDRAH 12
SUKRA 10
PIPRAWAL 3
KEWAI 14
DINGCHI 20
GAUHIA 17
BARWA 4
GHINHAL 5
GANTA 15

Fig. 7.3

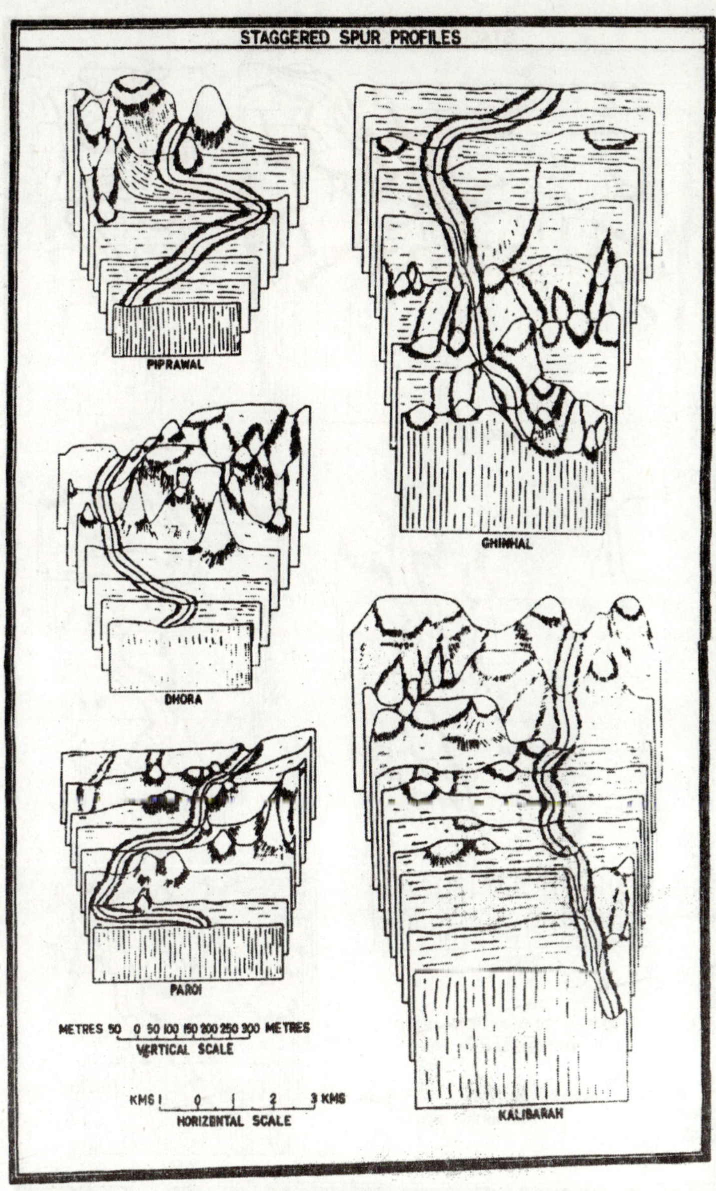
STAGGERED SPUR PROFILES
PIPRAWAL
GHINHAL
DHORA
PAROI
KALIBARAH
METRES 50 0 50 100 150 200 250 300 METRES
VERTICAL SCALE
KMS 1 0 1 2 3 KMS
HORIZONTAL SCALE

Fig. 7.4

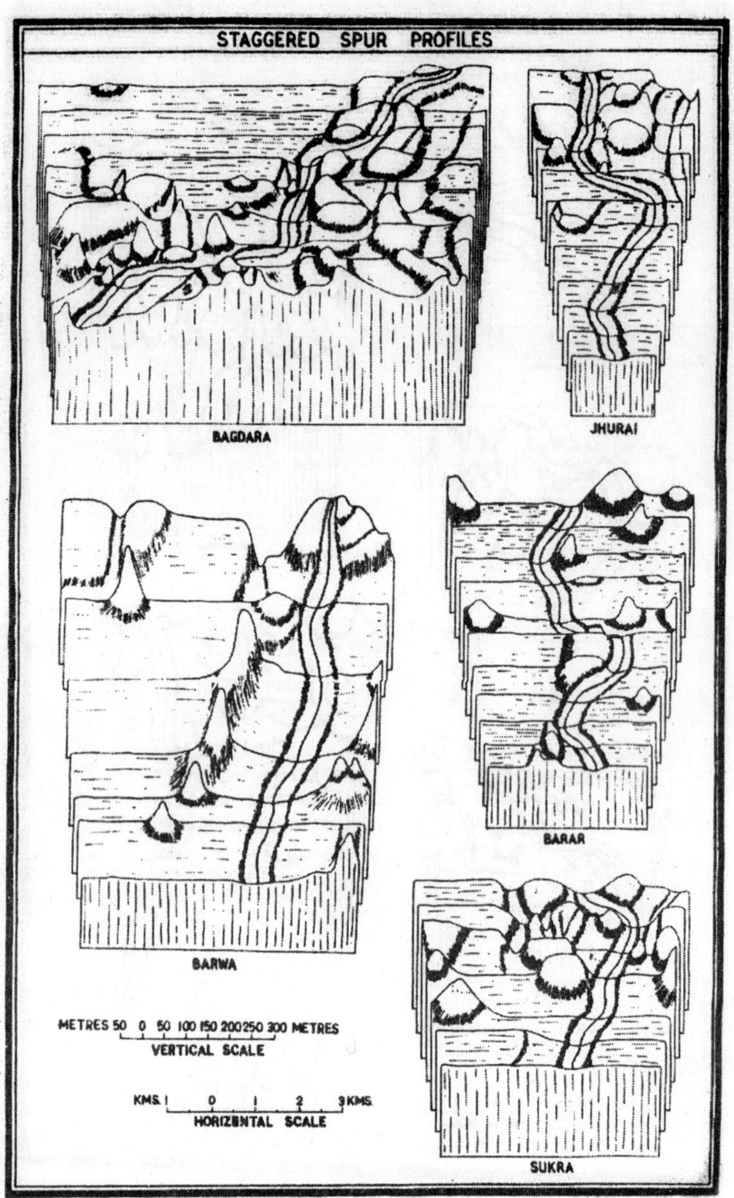
STAGGERED SPUR PROFILES
BAGDARA
JHURAI
BARWA
BARAR
SUKRA
METRES 50 0 50 100 150 200 250 300 METRES
VERTICAL SCALE
KMS. 1 0 1 2 3 KMS.
HORIZONTAL SCALE

Fig. 7.5

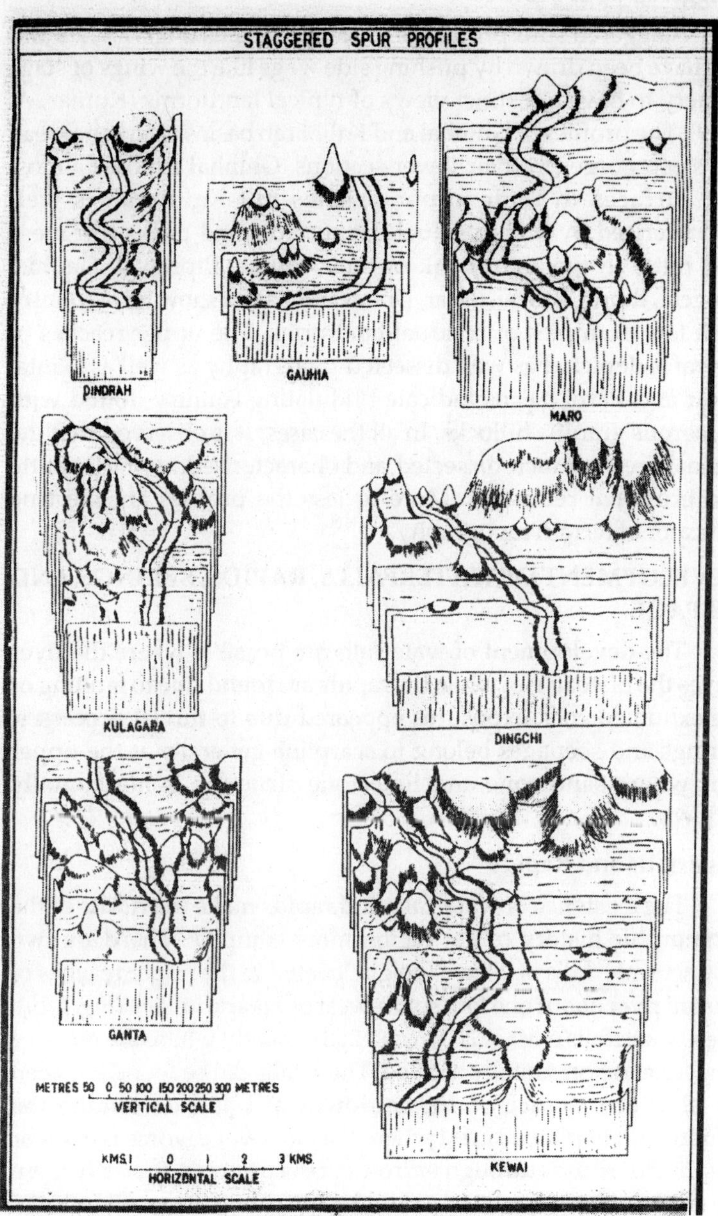

STAGGERED SPUR PROFILES
PINDRAH
GAUHIA
MARO
KULAGARA
DINGCHI
GANTA
KEWAI
METRES 50 0 50 100 150 200 250 300 METRES
VERTICAL SCALE
KMS 1 0 1 2 3 KMS
HORIZONTAL SCALE

Fig. 7.6

Staggered Spur Profiles

Staggered spur profiles of 17 sample basins (Figs. 7.4, 7.5 and 7.6) have been drawn by pushing side wags like the wings of stage scenery to have a distinct views of typical landforms (Kumar, A. 1979). The profiles of Ghinhal and Kalibarah basins clearly bespeak the steep gorges in their lower sections. Ghinhal exposes almost flat surfaces in their upper reaches but Kalibarah is well characterised by high residual flat surfaces and peaks like mesa and Butte in the right bank of their course. Piprawal, Gauhia, Dingchi, Jhurai, Kewai, Barar, Dhora and Barwa show level country with few examples of residual landforms. The upper reaches of Kewai is depicted as well dissected topography as well as Ganta. Barar and Paroi basins indicate undulating country dotted with numerous granitic hillocks. In all the cases, it is observed that the basinal area is much dissected and characterised by mesa, butte and erosional remnants. More or less the profiles are the true indicator of terrain topography.

DEVELOPMENT OF WATERFALLS, RAPIDS, SPRINGS AND SEEPAGE

The development of waterfalls are possible where the river leaves the plateau rim while the rapids are found due to faulting or the exhumed rock exposures appeared due to fluvial processes. Springs and seepages belong to scarpline generally at the upper part where sandstone and limestone structure is horizontally bedded.

Waterfalls and Rapids

The existence of waterfalls and rapids make the riddle of the geomorphic history of the region more complex. There are two distinct waterfalls of 20 m in height located in the upper reaches of Paisuni river about one km. south-west of Tikaria village (Fig. 1.15). These waterfalls exist at 280 m and 260 m height where the massive sandstone is horizontally bedded. These falls can be found between the shorter distance of 100 m downward and determine the termination of plateau crest where Paisuni river divorce the upper crest and descends through narrow cut deep gorges. Basically there are structural waterfalls in nature because there is no evidence of block upliftment or folding and faulting. Some may opine that the jerks caused by the Himalayan orogeny during Tertiary period might have caused relative upliftment in the northern foreland of India rejuvenating the existing streams. Based on this argument the Northern foreland as a whole should have been block uplifted and only the southern margin have been accentuated. But the numerous waterfalls in the inner part of the foreland like Chitrakut plateau and surrounding areas may be explained only when there is local upliftment in relation to surrounding area. But there is no evidence of any such local upliftment because the rock exposures along the detached hills and escarpments indicate perfect parallelism, Singh, S. and Pandey, R.S. (1983) have also stated similar views in the case of most adjoining region of Bhander plateau.

Following to geological cross section along new dam axis (II) of Paisuni dam project (Hukku, B.M. 1966), it may be recognised that there is two fault zone along the drill hole no. 12 and 19 (Fig. 1.13A). These faults show the shifting bed of Kaimur sandstone, Tirohan breccia and Tirohan limestone from top to bottom. This evidence denotes that the rapids as noted at Sitakund (Jankikund) about 1 m in height covering Tirohan breccia and limestone and Pramod Ban (about 1 m in height also over the similar structure, are undoubtedly caused by faulting. These rapids are continuously shifting behind because hard and soft limestone are horizontally bedded. Several micro-level landforms are formed over the upper skin of limestone due to solution of running water. Two waterfalls are also measured at different heights in the region. These are:

1. 30 m falls in the upper reaches of Ghinhal river (locally named as Chakra mala) at the height of 280 m and about 1 km north-west of Khadara village; and
2. 41 m falls along the tributary streams of Ganta river at the height of 180 m where the fingertip tributary leaves the western scarp face of Matdar Pahar.

A close observation of the waterfalls clearly reveals that these are neither real waterfalls as depicted on topographical maps of Survey of India nor heads of rejuvenation or knickpoints. Instead these are heads of embayments or scarp heads where waterfalls down the vertical walls during rains, but remain dry during rest of the year.

Two sites of seasonal waterfalls are also recognised along the tributary of Patraha nala 2-3 km north-west of Manikpur and along Allahabad-Satna railway line at the heights of 240 m and 220 m respectively. A spring condition is noticed near Bansachua which libricates the overlying sandstone country and ultimately leads to slumping and landslides. The differential weathering and the collapse of the rocks cause the condition of stepfalls. Another seasonal structural falls is noticed about 100 m away from Bansachua. These falls are 5 m to 7 m in height.

The position of falls are also validated by longitudinal profiles of the rivers (Fig. 7.2). The profile of Kalibarah (in the lower section), Ghinhal (in the upper reaches), Pindrah (in the lower reaches), Ganta (in the middle course). Ohan (in the upper reaches) and Paisuni (in the upper reaches) clearly indicate sudden fall in heights which are the causative factors of falls origin.

Springs and Seepage

Natural outflow of ground water on the surface is known as spring. There are numerous spring conditions and seepage sites, specially in those localities where sandstone and limestone strata are horizontally bedded (Fig. 1.15). The prominent springs in different localities are like Marpho hill spring (at 379 m height), Deo Pahar spring (at 180 m height), Khadara spring (at 320 m height in the upper catchment area of Ghinhal river), Gupt-Godavari springs (at 220 m in height), Madhwa springs (at 260 m height), Ram Nagar spring (at 320 m height along Bagdara basin), Retha spring (at 180 m height in the lower reaches of Jhurai basin), Hanumandhara spring (at 280 m height), Narsinghdhara spring (at 280m height), Devangna spring (at 270 m height), Kottirth spring (at 280 m height), Bankesidh spring (at 280 m height) and Bansachua spring (at 280m height) etc.

These springs are largely depended on a atmospheric precipitation and water bearing limestone structure overlying massive sandstone throughout the scarpline of the region. They are most common along the scarpline. The existence of seepage is also governed with limestone structure. Water appears on the surface where the country rock is weathered. These springs are the major source of water supply in plateau region because besides these water bodies there is no other source of water for residing population. It is remarkable that seepage sites are most common where spring conditions are noticed.

DEPOSITIONAL LANDFORMS

Depositional landforms are most common in the north plain along the lower section of the rivers. In the present region alluvial plain, fans and cones are notable landforms.

Alluvial Plain and Associated Landforms

The rivers which are partly or exclusively developed over flat and rolling surfaces of north alluvial plain, are well characterized by flat bottomed valleys with alluvium accumulation making flood plains. The food occurs during the rainy months and thus the major streams like Paisuni, Ohan and Ganta make flood plain along their lower courses. The river sudden dries after monsoon period and the transported materials along the river courses also become dry. The continuous sedimentations have caused the alluvium depth upto 50m in the lower section of the river valleys. The flood bottomed valleys are made because of lesser stream velocity, valley side slopes, channel gradient and comparatively the greater stream load in the northern plain of the region. The neighbouring countries of Yamuna-Paisuni confluence near Rajapur indicate the flat valley containing alluvium deposits along the both sides. The alluvium surface denotes mudflow and soil creep in wet condition and desiccation cracks in dry condition. The Paisuni river bed also makes alluvium terraces along their cross profile. Rivers like Paisuni, Ohan, Ganta and Banganga generally indicate meandering course in their middle and lower sections. Thus the sinuous flow path makes several pools and riffles in the region.

Alluvial Fans

Alluvial fans are the most important features of senile stage formed by river deposition in the lower section of foot hill zones. The river deposits the transported materials in a circular form followed by gentle slope (sudden break in slope angle) and low stream velocity.

A peculiar example of alluvial fans is seen near Sheorampur hill surroundings where the tributary streams has left the loose sand and morrum deposits in the lower section of the hill and formed a feature like alluvial fans. The sand deposits are not beneficial for agriculture because the grain size of the total deposited materials are the larger. The region experiences various examples of alluvial fans but these fans are important if they are made by major channel like Paisuni, Ohan or Banganga. If they are constructed by small tributaries generally in the rainy months, they produce sand accumulation problems for neighbouring agricultural lands because the small tributary streams have not found to contain water after rainy months.

Alluvial Cones

Alluvial cones are seen along the hill side slopes of the region due to deposition of transported materials of different sized and shaped. Alluvial cones exist by erosion and weathering processes. Weathering processes form talus cones due to mass translocation and deposition of rock wastes in the lower sections of the hills. The system of sedimented materials differ in each condition. Erosional cones exhibit big boulders while weathering cones (Talus cones) show finer rock particles in the upper part (foothill zones). The alluvial cones are observed to be well drained by drainage lines. The residual hills near Sheorampur and Bharatkup indicate towards the formation of alluvial cones where sedimented materials are sketched by parallel rills having rectilinear slopes. Most of the hillside area in the northern part of the upland denotes the formation of alluvial cones introduced by erosional and weathering processes.

CONCLUSION

On the basis of the above discussions, some interesting findings about the assumptions of landforms development can be tested following the hypotheses like:

Hypothesis-1

"That under prolonged period of stable environmental conditions and crustral stability, the landforms undergo progressive changes through time according to 'Davision model' of geomorphic cycle."

The present landscapes of 'Chitrakut upland' and the post-morphological history of the region clearly indicate the story of prolonged period of stable environmental conditions and crustal stability since Archean era to present time and thus the active denudational processes found a long period of geological history to change the impassive solid mass of Vindhya mountain into extensive peneplain of subdued relief.

Hypothesis-2

"That in certain circumstances, certain forms of landscapes do not change through time rather these are maintained constant according to dynamic equilibrium theory of open system."

The examples of several mesas having free face elements and scarp free faces truly speak the removal of weathered debris down the slope because of parallel retreat through the processes of backwasting. Although the top surfaces of the scarps are continuously diminishing back due to parallel retreat but their original form and steepness of slopes are still remain in equilibrium conditions.

Hypothesis-3

"That landforms remain in equilibrium so long as the internal and external environmental conditions are stable but for short-term period and if the stable conditions remain for long-term period, progressive changes in landscapes in the line of Davision model to occur leading to definite destination (planation surface)."

This hypothesis is a part and parcel of the aforesaid hypothesis and thus it also corresponds the reality as mentioned above.

Hypothesis-4

"That geological structure plays dominant role in the development of landforms so long as resistant formation rests over relatively less resistant formation."

The northern phases of the scarp are well characterized by limestone formation overlaid with hard and horizontally bedded sandstone rocks. The hard rocks expansion generally controls the top surfaces of mesas and scarps and thus the plateau maintains their original form by precipitus scarps. But where the hard rock cappings have been diminished due to prolonged period of parallel retreat these precipitus scarps and tabular topography have been converted into rounded shape denoting convexo-concave profiles. These rounded hills are being lowered little by little due to engagement of downwasting. Therefore, it may be stated that the present hypothesis gives more satisfactory results with reference to present landscapes of the region.

Hypothesis-5

"That morphometric variables reveal overall general characteristics of landforms components and these speak of disequilibrium if any caused due to recent upliftment".

The present hypothesis does not holds good in the case of Chitrakut upland because the dominance of low drainage density, low stream frequency, gentle slope, low hypsometric integrals, low relief ratios and longer length of overland flow clearly give the view of equilibrium stage of landform development. There are not a single process which can prove the disequilibrium stage and recent rejuvenation of the region.

Hypothesis-6

"That more than single factor operate in a region but a few of them emerge as dominant controlling factors."

In favour of this hypothesis, it may be stated that Climo-litho topo and tectonic factors play vital role regarding the evolution and development of present landforms scenery of the region.

Hypothesis-7

"That a region may experiences both backwasting (keeping landform in equilibrium due to parallel retreat) and downwasting (progressive lowering of relief and thus flattening of slope) at the same time depending on local conditions."

This hypothesis is perfectly validated in the present case because the region experiences backwasting of the scarps through the parallel retreat. The several isolated residual hills of northern alluvial plain are being progressively lowered due to downwasting processes. The examples of backwasting have been examined along the two different phases of 'Nilgarh pahar' in the north of scarp zones of the region. Thus the above fact strongly support the 'composite theory' of landform development.

All these factors validate that the assemblage of landforms can be explained with the help of the interpretation of processes variability.

8

Survey and Analysis of Some Major Landforms

Chitrakut and its adjoining area designates an assemblage of landforms. In the south, it is northerly dipping precipitous scarp running towards SW-NE directions. The scarpheads occupy a thin section of limestone strata overlying the Kaimur sandstone throughout the region. The limestone is dolomitic in nature and forms limestone cavern at different places wherein the Gupt-Godavari caves are peculiar in nature. The limestone formation is more or less characterised with spring conditions and seepage sites. The plateau is highly dissected towards north where downward slopes are ravaged by rills and gullies. The riverside alluvium deposits are also marked with ravines. The Yamuna plain and south plateau is attached together denoting detached residual hills along with granitic knolls at some extent. The granitic tors are the other significant features generally found along the downward slopes of northfacing scarps and along the residual hills.

Apart from the above, several micro-features are also noticed in the region but the significant features of the region are the ravincs (a detailed discussion will be given in Chapter-9) limestone caverns and granitic tors. In the present context, the micro-level study of some major landforms like limestone caves and granitic tors of the region have been taken into consideration.

INTRODUCTION TO SURVEY TECHNIQUES

Landforms may be studied with the help of field measurement, office work and laboratory-based analysis. Field measurement involves the extensive survey at micro-level. It can be also interpreted by topographical sheets and mathematical analysis in the office. A satisfactory result may be found by both ways e.g. taking a panoramic view of landforms by field knowledge and a mathematical design through office work. But laboratory work incorporates help partially from field measurement and partially from mathematical techniques. But the study of landforms by single technique is not fruitful. It is necessary to follow all the procedures at the time needed for landform measurement.

It is apparent that the nature and characteristics of morphogenetic landforms are already discussed earlier on the basis of extensive field measurement and the interpretation of the topographical sheets of the region. But the laboratory analysis is not included. Taking into consideration of the above points, the present investigator has applied the three way techniques of landforms' analysis. Micro-level survey has been done by repeated observations in the Gupt-Godavari area during 1982-1989. Similarly the granitic tors were also studied in Bharatpur and Sheorampur localities. The cave plan was prepared by Prismatic Compass survey. Air and water temperatures were recorded by precision thermometer. Chemical, micro-chemical and petrographic analysis of rock samples were undertaken in the Geological laboratory of Banaras Hindu University, Varanasi. Trace elements were determined spectrographically. The microscopic study of the rocks in thin sections was also undertaken. Overall materials are mechanised to interpret the landforms. The mode of formation, morphogenesis and the processes operating there are minutely discussed. It is also remarkable that the micro-analysis of the rocks confirms the history of the geological formation of the region. The findings are critically examined with the comparative knowledge of previous geological investigations.

MORPHOGENESIS OF GUPT-GODAVARI CAVES OF VINDHYAN DOLOMITE ROCKS

The caves exist as the significant features of the region. The two caves namely as 'Gupt-Godavari Caves' located at Gupt-Godavari hills are very popular since Ramayan period. These features attract through their natural beauty and historically developed as the religious sites. It is wonderful for common people but interesting for geologists and geomorphologists and morphologically remarkable in geological investigations. These two modes of interest in thinking attract the present investigator for micro-level study of Gupt-Godavari caves.

In this sense, the location, horizontal and vertical delimitation of associating geological stratigraphic sequence, mode of formation and geochemical analysis of associating rocks, plan, nature and characteristics of Gupt-Godavari caves, causes of solution and chemistry of rocks and morphogenesis of these two caves have been taken into consideration in detail. The three-way analysis (field work, office work and laboratory analysis) is incorporated wherein Prismatic compass survey plan of the caves is cartographically shown.

Location of Gupt-Godavari Caves

The Gupt-Godavari caves are located in the south-west of 'Chitrakut Dham' (Sitapur) at 25° 6′ N latitude and 80° 46′ E in longitude dolomitic structure of Vindhyan system (Fig. 8.1). The rock unit causing the caves is prominently known as 'Tirohan Limestone' (from the village Tarahuwan in Banda district) in the stratigraphical records (Prasad, C. 1986, 87, 88). Tirohan limestone belongs to Semri series and was first pointed out by Medlicot, H.B. (1989). It was later extended by Heron, A.M. (1922) to include similar limestone occurrences in Rajasthan area. The upper portion of Gupt-Godavari hills are layered with Kaimur sandstone. The rock is most soluble under karsting processes. The soluble dolomitic exposures are mainly observed between Lalapur in the east and Godha-Rampur in the West. It is also found in some of the isolated hillocks located in the north of the scarp-zone. Besides Gupt-Godavari caves, the dolomitic exposures of the area also denote few small cavity formations viz. at Modhwa hill (south of Brijlal Ka Purwa) and near Bankesidh (about 2 km south of Sidhpur village in the south-east of Karwi Town). These two caves are the attractive sites of 'Chitrakut Parikrama' from tourist point of view. It is an important feature geologically and is significant as a famous Hindu religious site.

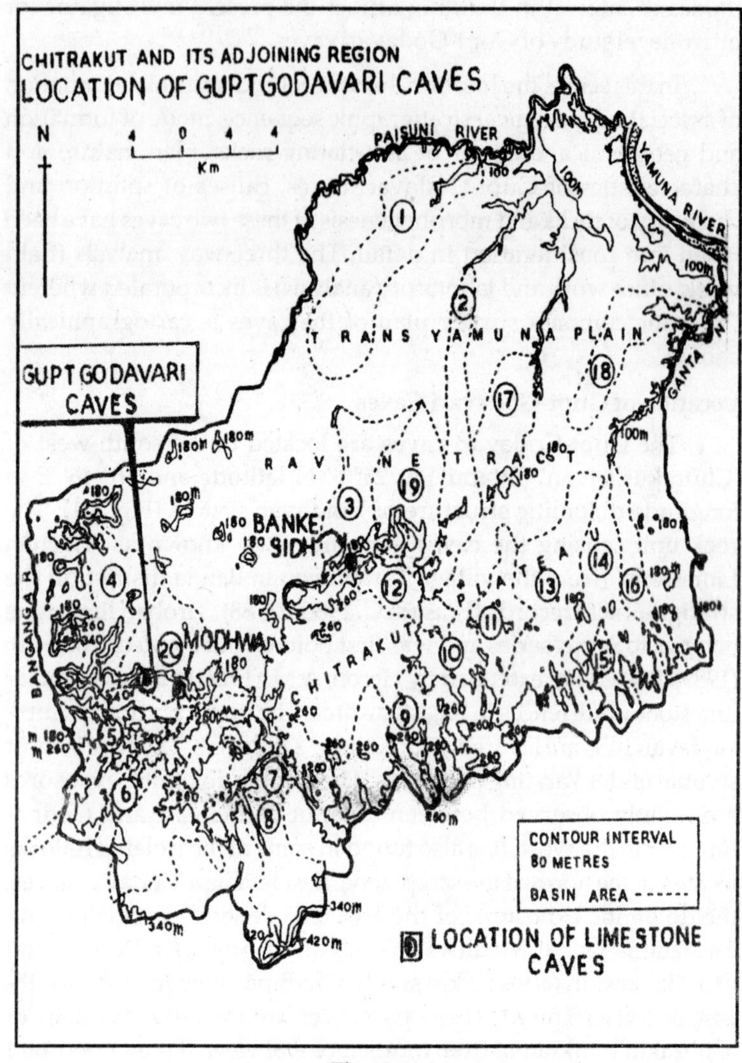
CHITRAKUT AND ITS ADJOINING REGION
LOCATION OF GUPTGODAVARI CAVES
N
4 0 4 4
Km
PAISUNI RIVER
YAMUNA RIVER
TRANS YAMUNA PLAIN
GANTA N
GUPT GODAVARI CAVES
RAVINE TRACT
BANKE SIDH
PLATEAU
MODHWA
CHITRAKUT
BANGANGA R
CONTOUR INTERVAL 80 METRES
BASIN AREA
LOCATION OF LIMESTONE CAVES

Fig. 8.1

Horizontal and Vertical Delimitation of Stratigraphic Sequence

In general, the thickness of dominate rocks in the area is about 60 metres. It occurs between 21m to 106m but the variation of thickness between 60 m and 63 m is most common. Direct section measurements (Safaya, H.L., 1963-66 and Hukku, B.M. 1963-66) have revealed the following thickness of dolomite rocks at different isolated hillocks:

Kamta Nath hill	—	61 metres
Lachhman pahar	—	20 metres
Sangrampur hill	—	60 metres

The delimitation of the original surficial extension of dolomite rocks is uncertain. It is extended towards south and south-east underneath the plateau. It seems to be greater in north-east and also appears below the alluvium cover.

Mode of Formation and Geo-chemical Analysis of Dolomite Rocks

(a) General Characteristics of Dolomite Rocks

The Vindhyan dolomite rocks are greyish to pinkish in colour. They are fine grained and massive. The specific gravity varies from 2.7 to 2.9 (Safaya, H.L. 1963-66). They are highly jointed wherein north-east and south-west directions are almost vertical. Examples of mud-cracks in the form of grooves are seen along the bedding planes. Shrock, R.R. (1948) defines that these grooves (mudcracks) are original features of limestone deposited generally on sloping surfaces.

(b) Dolomite and Associated Rocks

The dolomite rocks of the region is overlain by thickly bedded and brecciated chert (Tirohan Breccia) which is formed by primary precipitation of silica and chertification of the carbonates. Pockets of clay (mainly rich in kaolinite and illite), stromatolitic beds, pillet limestone, infraformational breccia, glauconitic sandstone and certain pellet limestone horizons rich in glauconitic sand are the associated rocks with dolomite.

(c) Environment of Deposition

Several pioneer workers like James, H.L. (1954), Ginsburg, R.N. (1955), Whitehouse, U.G. and Jeffrey, L.M. (1955), Pettijohn,

E.J. (1957), Carozzi, A.V. (1960), Twen Hofel, W.H. (1961), Cloud, P.E. (1962), Banerjee, I. (1964), Beales, F.W. (1965), Kahle, C.F. (1965) etc. have studied about the depositional characteristics of dolomite rocks and suggested that the said rocks were deposited in the sallow waters of a continental self.

(d) Source of Sediments

Fermor, L.L. (1906), Fox, C.S. (1928), Auden, J.B. (1933), Twen Hofel,. W.H. (1961) and Ahmed, F. (1962) have expressed their opinion about the Vindhyan basin and source of sediments deposited in the basinal area. In the above studies, it was suggested that the surrounding land of Vindhyan basin provided little material. Safaya, H.L. (1963-66) advocates that the bulk of the sediments responsible to the formation of dolomite was derived from an off-shore marine environment, carried by the diurnal spring and storm tides and deposited in the intertidal and supratidal environment.

(e) Model of Sedimentation

A group of geologists (as mentioned earlier) believe that the process of sedimentation in Vidhyan basin is much similar to the process occurred on the Andros islands (Bahamas) tidal flats, in the peninsula of Qutar in Persian gulf, on the island of Bonaire (Netherland Antilles), on the intertidal zones of modern carbonate deposits of Florida coastal areas, in the Mohawk valley, New York, U.S.A., near lake Ontario, U.S.A and Canada and near the cretaceous of Israel. It is also reported that the pelleted limestone formed in the intertidal zone while dolomite and stromatolitic horizons were deposited in the supratidal zone (mud flats) and algal flat environmental respectively.

(f) Mode of Formation

Limited information are available about the mode of formation of Vindhyan dolomite in the ancient geological records. Safaya, H.L. (1963-66) has stated that the presence of pellets and oolites within the Vindhyan dolomite and the stramatolitic structure are the major geological evidences which proved the broad scale dolomitisation (regional dolomitisation) operated with the mechanism of 'Capillary concentration.'

(g) Geochemistry of Dolomite Rocks

Based on the geo-chemical analysis of ten samples of dolomite rocks, the report of chemical composition is appended in Table 8.1. On the basis of the above report, it may be explained that CaO and MgO are the principal constituents of Tirohan limestone. MgO ranges between 22.835 and 25.506 while CaO varies from 47.493 to 56.082. The average of samples for these two constituents is 23.919 and 50.89 respectively. Another major constituents are SiO_2 (average 12.5111), Al_2O_3 (average 5.145) and Fe_2O_3 (average 2.156). The remaining constituents are positively lesser.

(h) Micro-chemical Tests

The micro-chemical test also indicates that the rock is mostly composed of dolomite having percentage between 80 and 95 (on the basis of point count analysis) in various samples. Powder samples and thin sections analysis denote that calcite and dolomite and most common in brownish portions while calcite is measured in white portions.

1. Petrography and Mineralogy

Tirohan limestone is fine grained wherein grain size varies between 0.005 mm and 0.018 mm. The sparry patches range in grain size from 0.02 to 0.03 mm. The pellets and oolites are mostly less than 0.02 mm in diameter and the pyrite cubes measure between 0.03 and 0.08. Desiccation cracks, bird's eye structures, pellets and oolites with fuzzy outlines are the common features. The calcite is formed in accicular form and the dolspars are scattered in the groundmass of neomorphic calcite. The other important minerals are quartz and feldspars which are mostly present in the groundmass. The iron minerals are also seen in the groundmass. The carbonate veinlets are composed of coarse mosaics of calcite and dolomite (brownish colour) while vug materials possess brownish coloured calcite and dolomite and white accicular calcite in an outer and inner lining respectively.

Table 8.1: Chemical Analysis of Chitrakut Dolomite Rocks (Wt. % Oxides Excluding H_2O^+ and CO_3)

Constituents	*Rock Samples*										*Average*
	I	*II*	*III*	*IV*	*V*	*VI*	*VII*	*VIII*	*IX*	*X*	
SiO_2	11.328	14.972	15.916	12.451	13.364	10.467	10.186	11.493	14.218	10.716	12.5111
TiO_2	0.425	0.484	0.571	0.571	0.580	0.375	0.629	0.541	0.545	0.705	0.552
Al_2O_3	5.382	5.640	4.654	4.637	5.284	5.626	4.905	3.473	6.815	5.185	5.145
Fe_2O_3	2.308	2.467	2.494	2.065	2.267	1.927	2.076	2.018	2.255	1.655	2.156
FeO	0.153	0.224	0.237	0.082	0.518	0.194	0.234	0.189	0.484	0.263	0.259
MnO	0.445	0.427	0.440	0.486	0.268	0.528	0.383	0.431	0.549	0.402	0.432
MgO	23.284	24.548	24.411	23.740	22.835	25.506	22.879	23.380	24.405	24.202	23.919
CaO	52.662	47.493	47.647	52.759	49.899	48.271	56.082	45.871	47.540	51.677	50.89
Na_2O	1.467	1.374	1.273	1.400	1.495	1.380	1.468	1.468	1.410	1.465	1.420
K_2O	2.115	1.904	1.892	1.467	1.462	2.140	1.609	1.501	1.453	2.288	1.771
P_2O	0.427	0.461	0.459	0.337	2.030	3.530	0.547	0.542	0.390	2.452	1.134
Total	99.996	99.994	99.994	99.995	100.002	99.944	100.998	100.007	100.074	100.003	100.189

Note: Rock samples tested in the laboratory, Dept. of Geology, B.H.U. Varanasai.

(j) Trace Elements Content

Trace elements were determined for five rock samples and the derived results are posted in Table 8.2. The average content of twelve elements have been also considered. The concentration of the trace elements is an outcome of the interplay between certain fixed cations during a subsequent period of diagenetic activity. The concentration indicates the positive evidence of life in them. The poor preservation was probably due to unfavourable environments and provide unsuitable conditions for the preservation of life.

Apart from the above explanation, the exploratory drilling log at hole nos. 7, 17, 22 and 23 (Fig. 1.12C), 17 and 18 (Fig. 1.12E), 11, 14, 15 and 19 (Fig. 1.13A), 18 (Fig. 1.13B.1) and 21 (Fig. 1.13B.2) clearly express the highly cavernous characteristics of Tirohan limestone. On the basis of the 350 rock samples, Safaya, H.L. (1963-66) has also proved that Tirohan limestone is dolomitic in character (about 82 to 95 per cent of dolomite content). The relationship between CaO/MgO rates and silica percentage as drawn after Safaya depicts the well defined general range of chemical composition of dolomite varying between 1.28 and 1.53. The geo-chemical analysis of 350 samples also indicates that the alumina and iron oxide percentage are essentially low while the acid insolubles contents have large variation.

Thus, on the basis of the field observation, laboratory analysis and consultation of ancient geological records, the following information regarding the nature and characteristics of dolomite rocks are discernable:

1. The average depth of dolomite rocks near Gupt-Godavari caves is 60 m below the capping of massive sandstone rocks.
2. The specific gravity varies from 2.7 to 2.9. The grain size of dolomite range between 0.006 and 0.018 while the sparry patches range in grain size from 0.02 to 0.03 mm. The pillets and oolites are mostly less than 0.02 mm in diameter and the pyrite cubes measures between 0.03 and 0.08 mm (Safaya, H.L. 1963-66).

3. The outer lining of the calcite and dolomite rocks in the caves are brownish in colour while and inner lining is of white accicular calcite (Safaya, H.L. 1963-66).
4. The two caves denote the horizontal bedding of the rocks wherein vertical and horizontal partings are visible in cave I. The second cave occurs massive rocks.
5. The cave I depicts theature like opening while cave II associates the zig-zag long and narrow tunnel.
6. The chemical analysis of the rock samples (Tables 8.1 and 8.2) and the chemical report of Safaya, H.L. (1963-66) indicate the maximum concentration of CaO and MgO in the rock and dolomitic character.
7. pH value is noticed as 8.30 in these two caves. The temperature varies between 24°C and 28°C in cave I during the month of March (water temperature 25°C and air temperature 24°C to 28°C). Cave II contains water temperature between 22°C and 26.5°C (24°C at entrance, 26.5°C in the middle flowpath and 22°C at the end) while air temperature differs between 24.5°C and 26.5°C. The temperature and humidity are more or less associated with seasonal change.
8. Cave I depicts circular to elongated pattern while cave II develops linear pattern of developing stages.

Plan of Gupt-Godavari Caves

The plan of Gupt-Godavari caves have been prepared on the basis of the Prismatic compass survey. The detailed description about the size and shape of the caves are given in Table 8.3.

Nature and Characteristics of Gupt-Godavari Caves

The size and shape geometry of these two caves differ while associating side by side. A general outlook about the nature and characteristics of these caves are systematically arranged in the following manner:

Table 8.2: Trace element content (in ppm) in Chitrakut Dolomite rocks

Elements	Rocks Samples					Average
	I	II	III	IV	V	
W^{+4}	30	30	15	10	5	18.00
Zr^{+4}	5	5	Tr	Tr	15	5.00
V^{+3}	10	15	20	20	25	18.00
Sc^{+3}	20	10	15	10	5	12.00
IN^{+3}	20	30	20	20	50	28.00
Y^{+3}	300	300	200	100	60	192.00
Cu^{+2}	10	30	40	30	20	26.00
Co^{+2}	5	Tr	Tr	5	5	3.00
Sr^{+2}	250	400	500	500	520	434.00
Ba^{+2}	120	210	100	120	100	130.00
Pb^{+2}	5	10	10	15	5	9.00
Li^{+}	15	5	Tr	–	–	3.00

Table 8.3: Cave plan (Gupt-Godavari Caves)

S. No.	General Characteristics	Cave I	Cave II
1.	Area (sq. m)	384	97
2.	Total length (m) according to selected stations	66.50	56.90
3.	Air length (m)	59.00	40.00
4.	Maximum height (m)	7.00	5.40
5.	Minimum height (m)	1.55	1.50
6.	Maximum width (m)	22.00	5.90
7.	Minimum width (m)	0.85	0.85
8.	Structure and bedding	Horizontal with growth framework	Stromatolitic bedding with fracture (F_R) porosity
9.	Shape	Elongated	Linear
10.	Hardness of rocks	Moderate and highly weathered	Highly compact

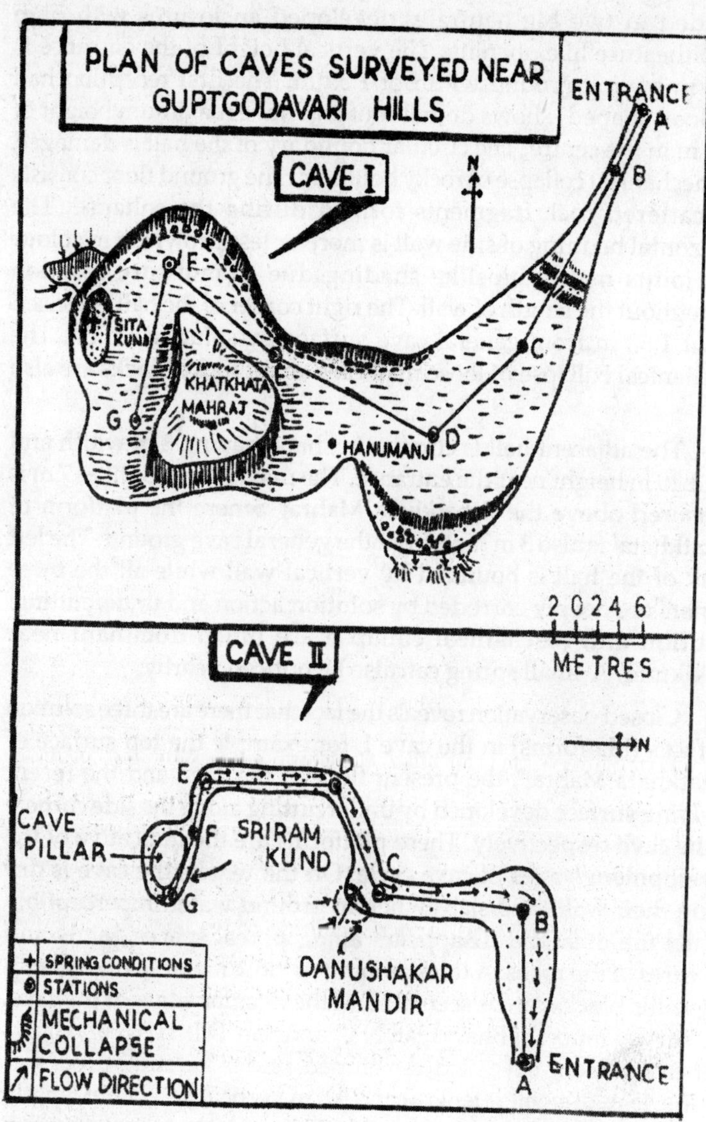
PLAN OF CAVES SURVEYED NEAR
GUPTGODAVARI HILLS
ENTRANCE
A
B
CAVE I
N
F
SITA KUND
E
C
KHATKHATA MAHRAJ
G
D
HANUMANJI
2 0 2 4 6
METRES
CAVE II
N
E
D
CAVE PILLAR
F
SRIRAM KUND
C
G
B
DANUSHAKAR MANDIR
SPRING CONDITIONS
STATIONS
MECHANICAL COLLAPSE
FLOW DIRECTION
ENTRANCE
A

Fig. 8.2

Nature and Characteristics of Cave I

In general, the cave is elongated in shape (Fig. 8.2A) and divided in two big naturally developed enclosures with high ampitheature like opening. The vertical height of the entrance is 1.55 m having ground width of 1.35 m. The first reception hall (balloon shaped) shows domal roof and have maximum height of 5.70 m in the centre. The circular boundary of the hall is damaged by mechanical collapse of rocky boulders. The ground floor consists of scattered rock fragments formed during the collapse. The horizontal bedding of side wall is more or less brownish in colour and joints make chipslike shading due to moisture reaction throughout the length of wall. The right corner of the hall is incised about 1.50 m from general cave surface by solution action. The mechanical collapse of loose materials along rocky blocks are also seen.

The adjacent hall is circular in shape having 8 m width and has 2.20 m height near the entrance. The maximum height as 7 m is measured above the 'Khatkhata Mahraj' where the platform of 'Khatkhata' is also 3 m high from the general cave ground. The left flank of the hall is bounded by vertical wall while all the three corners are deeply corroded by solution action and undercutting. Solution and mechanical collapse are much dominant near Jankikund. A small spring can also be noticed nearby.

Closed observation reveals the fact that there are three solution surfaces (Platforms) in the cave I, for example the top surface of 'Khatkhata Mahraj', the present floor of the cave and the recent lowlying surface developed by undercutting along the side corners of the cave respectively. These platforms are the indicative of the development stages of cave cycle. On the whole, the cave is dry along their walls and slabs wherein ground water imperceptibly drains the area and disappears along pores, spaces, joints and fractures of the rocks. A thick amount of loose materials along with dolomitic blocks can be seen around the circular edges of the cave. The survey report of Bhatnagar, N.C. and Sah, D.L. (1966-67) shows that the present cave was developed under karsting processes having 45 m diameter (length) and 4.5 m vertical height but length, height and width have been increased during 23 years of duration due to solution action and mechanical collapse. The greater variations have been noticed during the period of 1984 and 1986.

Recently a solution channel has developed about 5 m in length and 2 m in width near Jankikund by a suddenly collapse of the hanging roofs.

Nature and Characteristics of Cave II

The second cave is just like decorated arch of gallery wherein the entrance is designated in domal shape. It is elongated in shape having long flowpath of about 56.90 m length. It covers about 97 sq. m of solution area. The maximum height is observed at station C where the left corners of the cave is tended to progressive solution and mechanical collapse. The maximum height is surveyed between C and D stations. The maximum width (5.90 m) is seen in the reception room. The dolomitic rocks is fine grained and a massive pillar near Shriram Kund is also noticed. A hanging trimmings are notable along the cave roof resembling like stalactites. In fact, it is narrow-escaped hard dolomitic particles initiated by solution of spring water. The ground water seepage occurs round the year but the quantities of water run-off vary considerably subject to seasonal variations and precipitation in the area. The existing spring stream is embellished as the name and address of 'Gupt-Godavari' a famous site of 'Chitrakut Dham' for Hindu Pilgrimage.

Causes of Solution and Chemistry of Dolomite Rocks

The limestone rocks transport the two major compounds namely as calcium carbonate ($CaCO_3$) and dolomite [$Ca\,Mg\,(CO_3)_2$] is varying proportions. The little portion or totally lacking of dolomite makes the limestone rocks as calcic. The composition of calcium-magnesium carbonate and impurities with little calcium carbonate makes the rocks as dolomite. The limestone rocks of Chitrakut upland is dolomitic in character where calcium and magnesium contain several mineral content. Both the calcium and magnesium ions are absorbed by colloidal complex. Fresh water is not a active solvent but when it brings with a distinct amount of carbon dioxide (CO_2) it forms carbonic acid—

$$H_2O + CO_2 = H_2\,CO_3$$

(Pure water + gas = carbonic acid).

The carbonic acid or carbonated water attacks many rocks and is of very much dominating in the solution of limestone containing carbonate minerals—

$$Ca\,CO_3 + H_2\,CO_3 \rightarrow Ca\,(HCO_3)_2$$

(Calcite + carbonic acid → calcium hydrogen carbonate). Dolomite rocks which contain double carbonate of calcium and magnesium are more resistant than calcite ($CaCO_3$). The following reaction can be observed when the water containing carbonic acid attacks on dolomite rocks—

$$Ca\,Mg\,(CO_3)_2 + 2H_2O + 2CO_2 \rightarrow Ca\,(HCO_3)_2 + Mg\,(HCO_3)_2$$

If the calcium carbonate and dolomite are burned the reaction can be measured as follows—

$$CaCO_3 + Heat \rightarrow CaO + CO_2\uparrow$$

$$CaMg\,(CO_3)_2 + Heat \rightarrow CaO + MgO + 2CO_2\uparrow$$

The reactions involved in lime (CaO), hydrated or slaked lime Ca $(OH)_2$, and ground lime ($CaCO_3$) by carbonated water may be expressed in the following manner—

$$CaO + H_2O \rightarrow Ca\,(OH)_2 \quad ...(i)$$

$$Ca\,(OH)_2 + 2CO_2 \rightarrow Ca\,(HCO_3)_2 \quad ...(ii)$$

$$CaCO_3 + H_2O + CO_2 \rightarrow Ca\,(HCO_3)_2 \quad ...(iii)$$

The carbonation of calcium and magnesium hydroxides occur as follows—

$$Ca\,(OH)_2 + CO_2 \rightarrow Ca\,CO_3 + H_2O$$

$$Mg\,(OH)_2 + CO_2 \rightarrow Mg\,CO_3 + H_2O$$

$$\begin{cases} Ca\,CO_3 + 2HCl \rightarrow Ca\,Cl_2 + H_2O + CO_2\uparrow \\ Mg\,CO_3 + 2HCl \rightarrow Mg\,Cl_2 + H_2O + CO_2\uparrow \end{cases}$$

The pH value of a solution may be expressed as the logarithm of the reciprocal of hydrogen ion activity or concentration as given as—

$$pH = Log\,\frac{1}{(H^+)} = Log\,[H^+]$$

The presence of calcium and magnesium salts in dolomite rocks give a preponderance of H ions over H ions in the solution. The pH value of 8.30 measured in the caves water is considered as moderately alkaline. It shows moderate degree of reaction. The pH of solution depends upon relative amounts of adsorbed hydrogen and adsorbed metallic cations. The solution of caves tend to decline during the summer because of declining of pH value whereas increases in the winter and rainy months with increasing pH. It has been proved that the pH value decreases as the CO_2 concentration increases. The moderate pH (8.5) indicates the CO_2 concentration about 0.03 per cent same as in atmospheric air.

Temperature also causes the chemical reactions. It is estimated that the speed of chemical reactions is nearly doubled at the rise of temperature of about 10°C. The spring water and air temperature of the caves indicate moderate degree of reaction.

Morphogenesis of Caves

The caves attract the greatest attention of researchers, being the biggest among the ground water features. Rapid solution and chemical reactions are thought to be major factors involve to produce the caves while mechanical collapse is seen affecting the enlargement of the caves. The nature of cave formation and associated processes are related to the structure of the rocks and climatic conditions.

The two caves as located side by side represent two types of morphological features. In the case of cave I, it may be referred that the formation and development of the cave is highly governed with progressive solution above the water level wherein the active downcutting and thus mechanical collapse along the corner's cause continuous deepening and widening. The surface of the cave is seen to be lowered according to the downward movement of water table due to seasonal change. Thus the three solutional surfaces are formed in cave I with continuous lowering of water level. The graded floor is caused by spring stream as marked near Jankikund. The mechanical collapse of blocky fragments from the roof and edges of the wall are more or less moisture controlled weathering and separation initiated with spring water. It is also observed that the upper portion of the cave wall and roof are dry in nature. The solution, undercutting and breakdown of the materials are progressed along moisture zones. Thus, the development of the cave surfaces has been thought to be proceeded along spring water and so-called stream movement which were continuously lowered with downward collapse of associated rock mass. These surfaces also determine the stages of karst cycle.

The cave II is just like a tunnel. The narrow long route of the cave is well confined with Gupt-Godavari streams caused by existing spring condition. The water of diverted surface stream has deeply entrenched the soluble rocks by corrosion and developed an integrated cavern route above the running water surface. The arch-shaped gallery is caused by solution of soft dolomite rocks associating the hard, massive structure. Bhatnagar, N.C. and Sah, D.L. (1966-67) believe that the lower level tunnel is possibly formed along the joints by solution of the rocks of meteoric water. The solution of the rocks appear to be in the nature stage. The effect of an aggressive mechanical undercutting is observed near 'Dhanushakar Mandir'. The continuous undercutting and mechanical breakdown of the moistured rocky fragments (about 2 m in width) has occurred during 1980-85. The author believes that the continuous collapse will combine the two caves into one in near future. It is also remarkable that there is no sign of mechanical collapse except the site of 'Dhanushakar Mandir', the place which is associated with cave I.

Thus, the formation of these two caves validates the fact that the development and initiation of the caves took place along the ground water table while the enlargement has progressed above the ground water table due to mechanical collapse and breakdown of moistured zones of the caves. The general outlook of the cave is different which indicates that the 1st cave is principally moisture-controlled while the 2nd cave is resulted due to continuous solution and corrosion of draining spring stream. The size and shape of the caves are also caused by these two factors. Thus, it may be pertinent to point out that the formation of cave is not an easy process. It must be explained in the light of the structural position and associating complex karsting mechanism. The caves existed in the area are principally governed with spring conditions because the small cavities as located at Modhwa hill and near Bankesidh are developed along spring sites.

MORPHOLOGY AND EVOLUTION OF TORS

Tors is another significant feature which has been studied at microlevel. A special study has been considered in the north of the Vindhyan scarp generally in Bharatpur and Sheorampur areas where tors exists along the downslope of residual hills over granitic structure. The feature is closely associated with lithological occurrences and the size and shape of the tors are thought to be controlled under lithology and climatic conditions of the region. It is also reported that the surrounding area of Naraini is rich in granitic parent materials and, thus, the area denotes some peculiar examples of tors.

On the basis of the extensive field work and topographical interpretation of the nature, spatial distribution, general characteristics, classification, the associated lithological control assessment, causes of tors evolution and the evolution of tors evolution theories have been systematically arranged.

Definition of Tors

Tors, one of the most controversial landforms, are piles of broken and exposed masses of hard rocks particularly granites having a crown of rock blocks of different sizes on the top and clitters (trains of blocks) on the sides (Singh, S. 1977). Characteristically, tors are groups of spheroidally weathered rock boulders rounded in bed rock, which may be exposed as a basal rock surface or concealed by a waste mantle. Although sometimes confused with kopjes or stacks, the rounded shapes and the way in which the blocks are piled up or perched in rather improbable positions on the massive unweathered rock base (Faniran, A. 1971a) distinguish tors from these other landforms.

Spatial Distribution of Tors

Tors are usually found in the north of Vindhyan scarplines (Fig. 8.3) along the foothill zones and downward slope profiles of isolated residual hills characterized by granite parent basement. The size and shape of tors vary widely in different localities according to bed jointing pattern of the parent rocks and the nature and magnitude of subsequent chemical weathering. It is apparent that the geometry of joint pattern and the intensity of chemical weathering along with rainwash and solifluction give rise to various size and shape of tors. Different localities like Nilgarh Pahar, Kamadgiri, Parvat, Bihara hills etc. show some interesting features of tors. In the northern plain tors also exists along Itaura hill (along Piprawal nala near Itaura village), Gauri Pahar (near Chakaundh village), Lodhwara Pahar (near Lodhwara village in the right flank of Paisuni river), Surajkund and Jaresur Pahar (in the left flank of Paisuni), Satrajkifa hill, Agarhwa Pahar, Neolakhoh hill and Udara pahar where granitic exposures are extensively weathered at the surface.

In the south-eastern part of the region Simaria Charandasi hill (along Paroi nala) and Bhaunari hills (near Bhaunari village) are associated with tors. In the western part hill like Gonda ka Pahar, Chandra Maheshwari Gorra, Barra, Marpho and Hadiyar Pahar produce tors in the lower sections of the slope profiles.

General Characteristics and Classification of Tors

Tors may be posted at the top of the hills, on the flanks of the hills facing a river valley or on the flat basal platform ranging from 6 m to 30 m in height. The constituent blocks vary widely in size depending on the jointing pattern of the original rock and the type and the degree of subsequent chemical weathering. It is found that the most common sizes of tors range between 1 m and 3 m in diameter although block diameters reach about 5 m. It is apparent that the bigger block if settled one by one on the top of the others may constitute tors.

The shape and size of tors is closely associated with the geometry of jointing pattern of rocks. The widely spaced and tight joint system cause castellated tors as observed in an adjoining areas of Naraini. Massive blocks subdivided into arched sheets have initiated domical tors in the area. Elongated or tabular jointing form tabular tors while closely spaced joints have formed small rounded tors in the area. Apart from the above tors may be seen in cuboidal, globular and angular in shape associating with lithology and weathering impact.

On the basis of the size and shape geometry, the different categories of tors as recognised during field measurement are given as:

1. *Rounded Tors:* Rounded tors are seen over the lower section of Nilgarh Pahar. The rectilinear slope having boulders of different size and shape designates tors association. The ground and upper section of Pahari hills form rounded tors. The big size rounded granitic tors are seen along upward

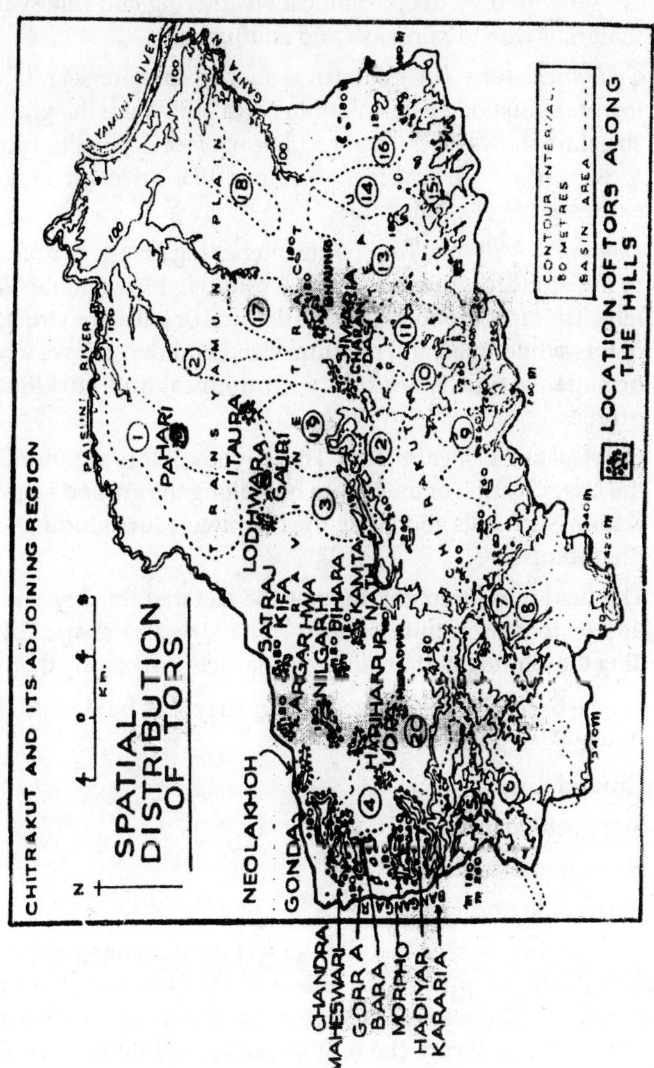
CHITRAKUT AND ITS ADJOINING REGION
N
0 4 8
Km.
SPATIAL DISTRIBUTION OF TORS
PAISUNI RIVER
YAMUNA RIVER
GANTA
TRANS YAMUNA PLAIN
PAHARI
ITAURA
LODHWARA
GAURI
SATRAJ
KIFA
ARGARHWA
NILGARH
BIHARA
KAMTA NATH
HARIHARPUR
UDRA
SIMARIA
CHARAN DAS
BHAUNRI
NEOLAKHOH
GONDA
CHANDRA
MAHESWARI
GORRA
BARA
MORPHO
HADIYAR
KARARIA
BANGANGA
CONTOUR INTERVAL 80 METRES
BASIN AREA
LOCATION OF TORS ALONG THE HILLS

Fig. 8.3

convexity while the transportational mid slope and ground slope register comparatively small rounded blocks. Some boulders are partially exposed over the ground surface. The tors field indicate globular jointing in granitic rocks which are separated by deep chemical weathering and removal of materials due to rainwash and solifluction.

2. *Cylindrical Tors:* The cylindrical shaped tors are seen in the lower section of western Kamta Nath hill where the granitic structure shows blocky and rectilinear jointing. Granitic blocks are observed vertical in nature just like tower. It may be expressed as elongated tors.
3. *Domical or Millstone Tors:* Some interesting results of domical or millstone tors are seen in the western flank of Kamta Nath hills. The big size boulders make domical or millstone structure of tors while small and medium sized boulders show cupola or sugar loaves types tors with domical and curvilinear profiles.
4. *Cuboidal and Elongated Tors:* These types of tors are found in the lower section of the Bihara hills along the ground slope of Kamta Nath hills and along the hill ranges surroundings the Bharatkup.

The field measurement denotes the fact that the area shows all kinds of tors in granite country. The size and shape differ according the jointing pattern and rock particle associating therein.

On the basis of the existence of tors, they may be classified in the following categories:

1. Ground tors;
2. Slope settled tors;
3. Hill top tors; and
4. Valley side tors.

Ground tors are seen in the lower section of Kamadgiri Parvat and along the Bharatpur-Bihara hills. The rectilinear slope profile of Bharatpur hills indicate slope settled tors. The tors located on the top of the Pahari hills are the best examples of hill top tors. But valley side tors are not seen in the area.

On the basis of the geological structure lithological control and the impact of weathering processes, tors of the region, may also be grouped as:

1. Rounded tors;
2. Millstone tors;
3. Cylindrical or tower like tors; and
4. Elongated or tabular tors as discussed earlier.

Tors and Associated Lithological Control

As a matter of fact, tors may be found in granitic country in different climatic zones. Tors like features are also seen in most of the area underlained by sandstone rocks but they are not real tors. The granitic country which is the parent rock of the area is the major source region of tors formation. The differential weathering and erosion have extensively caused the granitic rocks associating with tors of different size and shape. Thus, the granitic cover provides the basis for tors existence in the region.

Causes of Evolution and Evaluation of Development Theories

The origin of tors has been thought as the subject of much controversy. Various theories like two stage process theory (rottening of rock by chemical weathering and exhumation of rock decay by running water) of Lilton, D.L. (1955), Periglacial theory of Palmer, J. and Neilson, R.A. (1962) and Pediplanation theory of King, L.C. (1958) etc. have been put forth but there is no unanimity among the experiments. The observation of tors and associating features are not confined with a particular rock type and climate but a variety of rocks and climates claim their existence. But it is obvious that most of the tors are found to be formed over granite rocks having felspar minerals in abundance and the hardness, jointing and susceptibility to block and granular disintegration of granite offer ideal conditions for their formation.

Linton, D.L. (1955) has suggested that tors are resulted from sub-surface (rather than subaerial) rotting of granite through the action acidulated rainwater penetrating along the joints into the body of the granitic mass. On the basis of the observation of tors of the region, the following stages may be suggested for their evolution:

1. The existing Bundelkhand granite has developed incipient joints because of tensile tresses caused due to the contraction during cooling. The incipient joints must have been of two types: (a) joints having rectangular pattern; and (b) pseudo-bedding joints characterised by close spaced and curvilinear joints. These joints were further accentuated by shearing and tensional forces occurred during the period of crustal movement.
2. The continuous weathering and erosional processes (chemical weathering assisted by acidulated rain water divided the granitic mass into cuboidal blocks) resulted the degradation of the region culminating into peneplain and exhumation of hard and resistant granitic domes. It is notable that the deep basal weathering as postulated by Lilton cannot be ignored. The rocks of older granite were rotted to a greater depth and the materials along the joints and fractures were removed by denudational processes. The above removal, rather exhumation have caused the dilation of granite mass due to pressure release leading to recoil of rocks upward and thus the joint-diseased were exposed to further attack.
3. Following the gelogical evidence during the period of Tertiary upliftment, it may be stated that the upliftment may be held responsible for enlarging and further accentuating the pre-existing joints resulting into block disintegration of granitic domes. Continuous process of chemical and mechanical weathering and removal of debris by rainwash have caused the formation of granitic tors in the following manner:
   (a) Chemical attack along the joints of vertical and horizontal pseudo-bedding converted the granitic block in cuboidal structure.
   (b) Mechanical weathering assisted by slow decomposition of granite rocks having widely spaced jointing has designated in round shaped blocks.
   (c) The areas having curvilinear sheet jointing at the surface created by pressure release (as seen over Bihara hill) might have resulted angular shaped blocks.

(d) Onion weathering (exfoliation weathering) caused by interactions of mechanical and chemical weathering might have produced the removal of outer shells one after another and further furnished angular to rounded blocks of granites. Rainwash has played vital role removing the commuted materials and leaving the granite blocks resting as one upon another.

(e) The impact of weathering along the strong lateral fractures have caused tabular blocks.

(f) Weathering along the wide spaced hard granite joints have produced millstone blocks near Kamta Nath hill.

(g) Weathering agents initiate vertical blocks and vertical fractures.

CONCLUSION

The microlevel study of limestone cavern and granitic tors has revealed the fact that the region is principally governed and controlled with sub-surface and surface processes. The cave formation confirms the intensity of solution in karst area while tors indicates towards the deep weathering in Bundelkhand granite, the parent rocks resting at varying depths overlying alluvium blanketing in the region. The solution of the compact dolomite and the separation of granitic block according to their joint pattern clearly highlight the longer geological history and the later mature or senile stage of terrain development.

The landforms and existing peculiarities are more or less previously based upon litho-structure and later on operating processes with a long passage of time.

9

Environmental Problems

INTRODUCTION

Everything in environmental history is changing in nature. One of the greatest problems facing mankind today is to achieve a nice balance of nature between a rapidly increasing population and human thrust. Land degradation and the loss of fertile top soil have challenged the human race while techno-scientific encroachments have depleted the natural resource base of this planet. Thus, the wholesale cutting and ravaging of our forests, the removal of native sod cover, the senseless overgrazing, unnatural drainage operations and agricultural techniques, drought damage, floods, dust storms etc. are the major problems alarming the need of environmental improvement.

TYPES OF ECO-PROBLEMS

History denotes that the Chitrakut and its adjoining region was virtually a cockpit of constant warfare among the regional power like and Bundela rulers (e.g. the Marathas and the Rohillas) from the beginning of the 13th to the end of the 18th century. Due to warfare, the agriculture was totally neglected and scarcities and famines of that time occasionally deepened the crisis. A large number of tanks built by Channels, the canals by Britishers and the various projects and land reform measures since independence were the remedial processes accepted in the region with the passage of time (Saxena, J.P. and Lal, K.K. 1977).

After Independence, the extension of agricultural practices have almost removed the original forest cover. The dense mesh of drainage lines and the thick alluvium deposits have subjected to degrade under gullying and sheet erosion generally in the northern part. The unfavourable climatic conditions and the traditional cropping pattern are responsible for a meagre subsistence economy in the region. Irrigational facilities are rare while transport and communications are also found lacking.

The upland region is highly dissected having incised deep valleys of seasonal torrents. The forest covers are mostly reserved and are sparsely populated. The sandstone areas produce 'Patha' soils which is poor and unconductive for agriculture.

Apart from the above, sand accumulation in agricultural fields, rocky wastes and thorny scrubs of barren terrain many of the associated problems. The most dominating problems are deforestation and gully erosion. The stone quarries and water scarcity have aggravated the problem further. The soil erosion is thought to be the greatest geomorphological hazard.

Deforestation Problems

Forest as a renewable resource contributes goods (mainly food and fuel) and services to the people of the region. The forest culture of the Chitrakut is richly illustrated during 'Ramayana period'. The forest destruction and denudation principally governed by population explosion and livestock leading to enhance requirement of food, timber, fuel, wood and grazing respectively are the crux of the problem. The existing tribal population and local communities have misused the forest resources.

The northern plain area is dominated by scattered trees while open scrubs and fairly dense mixed forest are observed on residual hills and the southern plateau respectively (Fig. 1.17). There are about 41 reserved and protected forests wherein 16 (e.g. western Padari 13, south Padari 14, Hanumandhara 21, Duzurg 22, Dadari 23, Gursari 26, Kihuniyan 27, Parwania 29, Judehi 30, Anusuiya 31, Gupt-Godavari 33, Bharatput 36, Hariharpur 37, and Pokhai 41 wherein numbers indicate the serial numbers of the forests on Fig. 1.17) forests are protected and remaining 25 are the reserved. The open mixed forest located in the south and south-eastern part of the region, (Satna District) is degraded due to population pressure. Partial deforestation is also noticed in the surroundings of Sitapur,

Kamtanath temple, Hanumandhara and Gupt-Godavari hills. Overgrazing and clearing of the forest is most common along the residual hills of northern plain. The ravine tract also denotes forest degradation due to gullying (Plate 40). The forests are generally cleared for agricultural purposes and domestic uses such as wood, fuel etc. The forest area of the region is also mismanaged because the U.P. and M.P. Governments have not taken satisfactory actions for their preservation and improvement. It is also notable that the forest area of the region is overgrazed while they are reserved or protected in nature. Thus, the forest rules are violated on a large scale. Slope cultivation processes have caused the damage of vegetal cover along the most of the residual hills of the region.

The another cause of deforestation is the construction of hill roads in the area. It is remarkable that the tourists's interest sites (generally Parikrama Marg) like Sati Anusuiya, Kamtanath, Hanumandhara, Gupt-Godavari, Bharatkup, Lakshman Pahari, Pramodban, Purani Lanka, Jankikund and Sphatikshila are now well cemented by metalled roads. These holy places have added many infrastructural facilities like Dharamshalas, recreation sites etc. near the old temples. These activities have finally caused the clearance of forest land of the region. There has been a serious impact of small scale industries mainly quarring, toy-making and 'Biri' making in the area.

The stone quarries have raised serious problem of soil erosion, and mass movement leading to deforestation. The construction of railway lines and engineering works have also accelerated the deforestation.

Problems of Stone Quarries

The problems caused by stone quarries are common throughout the region. The area is rich in an excellent quality of Kaimur Sandstone, Tirohan limestone and 'Kanker'. The workable stone lies in beds from 15 cm to 3 mts. in thickness. Stone quarries are generally located in the north of the scarplines. The limestone quarries are seen near Gupt-Godavari and Khoh pahari. Sandstone and morrum quarries are in operation near Sheorampur area. It is apparent that these quarries accelerate erosional capacity initiating various forms of mass movement. Most of the isolated residual hills are degraded by quarries.

Problems Relating to Soil Moisture and Water Deficit in Different Litho-Units

The Chitrakut region is thought to be as a water scarcity region. The existing soils have shown low moisture content and ground water differs regionally on the basis of rocks and soils structure. The interlocking grains of sandstone rocks have no pores and spaces and thus the ground water occurs and flows along the joints, shear zones and other similar opening of secondary origin. Sedimentary rocks (alluvium cover) are characterised by presence of mineral grains and traversed by joints, shear zones and several zones of weakness which provide suitable conditions for free percolation of ground water (Prasad, G. 1986-87).

There are mainly four types of water bearing formations in the areas.

1. Recent to sub-recent alluvium comprising clay, silt, sand and gravel with subordinate quantities of kanker.
2. Weather granitic residuum comprising fragments of granite, quartz, felspar etc. and the jointed granite.
3. Cavernous and jointed limestone.
4. Jointed Kaimur sandstone.

The alluvium and weathered granitic residuum show similar characteristics of water holding capacity. In alluvium country the zone of saturation is chiefly formed by clay in silt with varying depth of water level (Fig. 1.18A and C) ranging less than 3 m to about 20 m below the ground level. The maximum depth (21 m) of water level has been recorded in some localised areas and the maximum depth of the open wells (24 m) is recorded in this tract ending to the clay. The depth of water appears in increasing order towards the Vindhyan scarps in the eastern part of the region. The water levels are subjected to seasonal variations and the yields from the wells are generally poor to moderate. It is also observed that the wells generally dry up within one hour of pumping and the complete recuperation occurs in 24 hours. This is the reason why the alluvial country around Karwi does not seem to be capable of supporting the heavy duty tubewells while the tubewells in the buried channel have better chance to provide good quantities of water. On the basis of the Geophysical survey of ground water survey department, it is noticed that few sites may possibly yield sufficient quantities of water through shallow tubewells. The above discussion highlights the fact that these two lithological units have not retained sufficient soil moisture and water deficit problem is most common in them.

The outcrop of cavernous limestone is massive and well joined wherein the permeability is mainly of secondary origin. It is highly cavernous in nature which is clearly exhibited by Gupt-Godavari caves. The Gupt-Godavari caves indicate heavy seepage of ground water and about 10 causes of Paisuni water is reported to be lost through open spaces and channels in the limestone around Jankikund. The water level in the Hanumandhara is well rested at 26 m below the ground level. It surges upwards by 12 m to 16 m due to heavy rainfall in the area. Following the well hydrology (Figs. 1.18B, 1.19B and 1.20B) it may be said that the circulation and the volume of water available to the wells in the limestones are depended upon the source of recharge of solution channels and joints of rock.

In general, there is perennial source of recharge of ground water in limestone and the associated soil possesses sufficient moisture and good quality and quantity of water. The major part of rocky terrain is characterised by Kaimur sandstone. Ground water occurs between the joints and the bedding planes of the poorly bedded rocks as well as in thin siltstone layers in the sandstone. Numerous open wells to depths as deep as 61.80 m below ground level have been excavated in the sandstone. The depth of water level (Fig. 1.18C) generally ranges between 2.5 m and 57.47 m below the ground level in the plateau region. The depth of water ranges from 18 m (Fig. 1.18A) to as much as 57.47 m in Ohan catchment area, 12 m in the Markundi-Manikpur tract 4.62 m to 8.62 m around Baraunda, 21 m to 30 m near Kichchari, 8m around Nadwania and 3.5 m in Kohilehie open wells respectively. Generally, it may be said that the Kaimur sandstone may not sustain even moderate supply of water through tubewells. Large diameter open wells with infiltration galleries at suitable sites, may yield moderate quantities of water.

Due to lack of sufficient ground water quantity, the region experiences the shortage of irrigational facilities and the supply of adequate water. The dry rocks and associating soils thus not suitable for agricultural crops. The problem of drinking water becomes more serious in the summer throughout the region.

Soil Erosion and Natural Hazard

Soil erosion which has comprehensively upset the ecological balance and ravaged the lithological regimes is boldly speaking as the rape of the earth. As far as the earth scientists are concerned, the chief culprit of soil erosion is due to dynamic change in climatic conditions, structural changes and several abnormal processes supported by man's activities like deforestation, land ploughing, overgrazing, overcropping, faulty irrigation and drainage schemes, quarrying, mining, urban development, road construction, forestry, recreation etc. respectively.

Soil erosion is the detachment and removal of the skin of the top soil either partially or completely. The finer portions and the fractions of highest fertility always removed first in soil erosion. Soil erosion and ravination has created the greatest problem and disturbed the nice natural balance of Chitrakut region.

Factors Affecting Soil Erosion

Soil erosion primarily results from physical causes and human mismanagement. In the present study region, soil erosion has been reached to serious dimensions due to physico-anthropogenic factors with the passage of time. The major factors controlling the soil erosion are as follows:

(a) Physical Factors

1. *Geological Structure:* The study area contains the geological formation starting from the oldest Archeans upto recent alluvium. The northern plain is deposited by alluvium while upland country denotes Kaimur sandstone and Tirohan limestone. The sandstone area is partially affected by soil erosion but the limestone structure accelerates incipient rills and progressive gullies due to soluble character. The alluvial plain is thought to be underlain by medium to very coarse sand and gravel with clay, kanker and silt. Gullying occurs

along the river courses due to easy erodibility of the formation. The soft theature of alluvium is characterised by presence of kanker pan below certain depth. The depth of gullies principally depend on the layer of the kanker found below the clay sand and gravel. Kanker plays vital role in the formation of rills and gullies through swallow holes and tunnelling in alluvium area.

2. *Rainfall intensity:* The area experiences sub-tropical climatic conditions wherein prickly hot summer and near freezing winters cause heating and cooling of the rocks which in turn creates insolation weathering. The mean annual temperature of 26.02°C and mean annual rainfall of 96.65 cm of the region support to maximum intensity of fluvial erosion according to Peltier's and Wilson's concepts. The summer rain, unless it is extremely heavy or persists for a long duration causes much destruction. One bare surfaces, the impact of torrential rain causes the detachment of soil. The removal of soil particles occur due to rain drop action wherein beating of raindrops destroy granulation while their splashing causes considerable amount of transportation. Thus the soil erosion takes place during the rainy months due to excessive run-off of existing streams on sloping as well as undulating country.

3. *Floral Forms:* The maintenance of an adequate floral forms is the surest way of preserving the soil and preventing the soil loss. It provides a natural protection for soil and prevent and check the soil erosion. As regards the present investigation, it is apparent that the upland country is covered with dense mixed forest and less consumed by erosional agents while the virgin barren land on residual hills dotted with thorny scrubs and grasses exhibit the serious impact of rilling and ravination. The land associated with major river courses is subjected to pipe flow and gullying due to clearance of vegetal shading.

4. *Pedological Characteristics:* On the basis of the assumption of pedological characteristics, there are three major soil categories on the upland. The upland soil which covers about 2/3 land area is highly susceptible to erosion. The sloping country towards the northfacing of the scarps is characterised by two

major soil groups e.g. Pedocals (Ca) and Pedalfers (Al + Fe) and encourages the rilling and ravination. Lowland and riverine soils are formed by alluvium wherein erosion occurs along the river courses but the depth of gullies depend on the thickness of alluvium deposits.

5. *Terrain Topography and Slope:* Topography and slope confirm the soil erosion. The flattop on upland is marked by sheeting, while steep slopes of the scarp denote maximum erosion. The lower sections (Terminal points) and the undulating country having gentle slope are seriously plaughed by ravines (Fig. 9.1). Rills are generally associated with sloping ground. In general, it is believed that the greater the degree of slope, greater the degree of erosion.

(b) Anthropogenic Factors

Stone-excavation and deforestation are the major factors caused by human activities. Collapse of rocks and mass movement are the geomorphic evidences which are also processed due to human mismanagement. Faulty irrigation, inadequate contour ploughing and engineering works are also responsible for soil erosion in the region.

Types of Soil Erosion and Its Intensity

There are many types of soil erosion but geological erosion and accelerated erosion relatively have its greater importance. The following major forms of soil erosion are recognised in the area:

(a) Sheet Erosion;

(b) Gully Erosion; and

(c) Cataract Erosion (Vertical Erosion).

(a) *Sheet Erosion:* Uniform skimming off the top soil with every extremely heavy rain persisting for a short duration or a long period is known as sheet erosion. It generally occurs over the top surfaces of the plateau in the southern upland, bare rock tops between residual isolated hills and along fairly gently sloping ground of the northern plain (Fig. 9.1). The swelling of the land surface by continuous sheet wash encourages the removal of a thin upper layer of soil in the corresponding area. In the plain area, the land mostly away from the river courses are closely associated with sheet erosion.

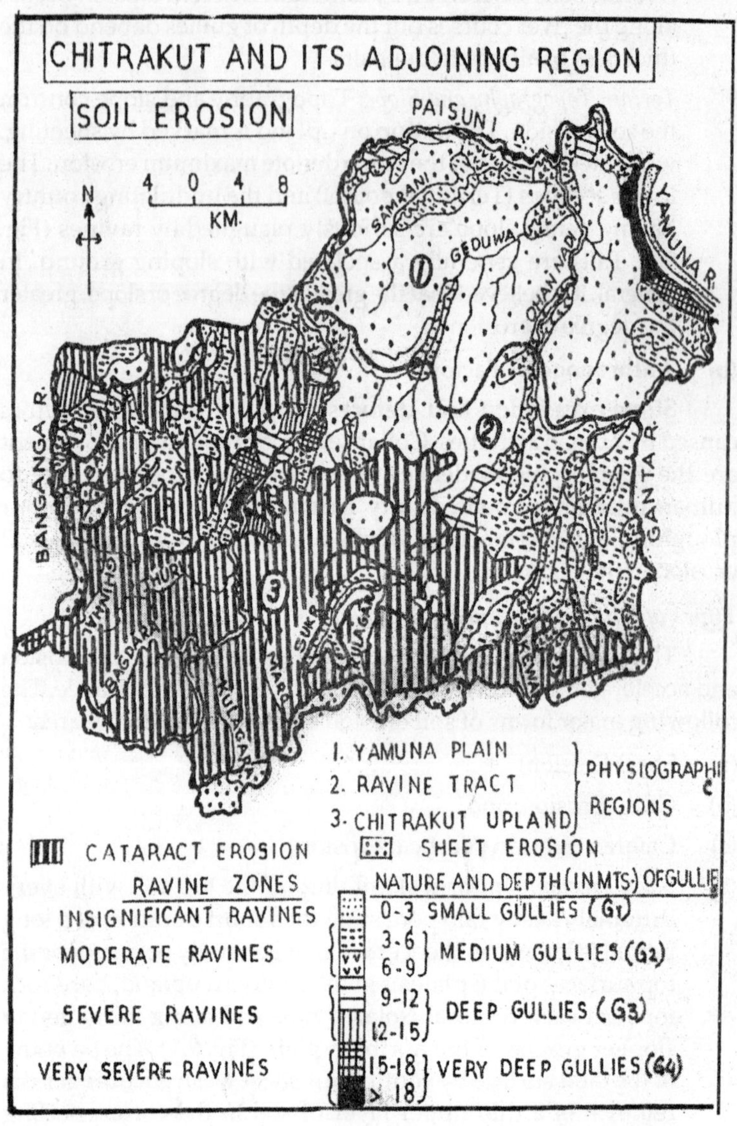
CHITRAKUT AND ITS ADJOINING REGION
SOIL EROSION
PAISUNI R.
N
4 0 4 8
KM.
SAMRANI
GEDUWA
OHAN NADI
YAMUNA R.
BANGANGA R.
GANTA R.
SUKR
MARG
BAGDARA
1. YAMUNA PLAIN
2. RAVINE TRACT
3. CHITRAKUT UPLAND
PHYSIOGRAPHIC REGIONS
CATARACT EROSION
SHEET EROSION
RAVINE ZONES
NATURE AND DEPTH (IN MTS) OF GULLIES
INSIGNIFICANT RAVINES
0-3 SMALL GULLIES (G1)
MODERATE RAVINES
3-6
6-9
MEDIUM GULLIES (G2)
SEVERE RAVINES
9-12
12-15
DEEP GULLIES (G3)
VERY SEVERE RAVINES
15-18
> 18
VERY DEEP GULLIES (G4)

Fig. 9.1

(b) *Gully Erosion:* It is chiefly produced by the excessive run-off of the major transporting agencies on the sloping ground along the north facing scarps as well as along the bank of major streams of northern lowlying country. Gullying is the progressive stage of sheet erosion. Sheet erosion initiates rills and deepening of rills have caused gullies. Rills are the intermediate state between the sheeting and gullying. Rills develop along the sloping surface when running water tend to concentrate into small rivulets soon after divorcing the hill crest. Gully and ravine erosion (Fig. 9.1) is most prominent in the present study region. There are some differences between gully and ravines specially in their morphological and genetic characteristics.

The word 'ravine' denotes gullied land containing systems of gullies running more or less parallel to each other and entering a nearby river flowing much lower than the surrounding table-lands. Generally, gully initiates along the animal trails, roads and path on the agricultural uplands. On the contrary, the ravine begins along the riversides and encroaches the catchment area by headward growth. Hence ravine is said to be a form produced by river action and gully as the function of catchment area.

The close observation of the region reveals the fact that the wetted zones whether it is in the lower sections or along the hill slopes are seriously ravaged by ravination processes. The hill top having gentle slope and the updulating surface of Paisuni-Yamuna plain is depicted by sheet erosion.

(c) *Cataract Erosion:* The southern plateau region is characterised with varying relief features, slope complexity and moderate to high dissection. The major drainage lines like Paisuni, Ganta and Ohan make gorges through vertical erosion in their upper reaches. The gorge position of Paisuni along their upper reaches is significant and notable in respect of cataract erosion. Ghinhal, Bagdara, Kulagara, Maro and Sukra river basins also develop vertical gorges partially along their upper courses. Thus, the crisis of cataract erosion is most common along the upper reaches of most of the drainage lines of the plateau region (Fig. 9.1). The convexo-rectilinear slope of the scarp is principally governed with cataract erosion. The lower sections of rectilinear slopes are generally ravaged by rills and ravines. Sometimes it appears that the area around the cataract erosion is complex in nature where rilling, sheeting and ravaging are partially marked. Thus, the erosional complexity becomes difficult to recognise in this sense. In general, the terrain above the height of 200 m falls under the effect of cataract erosion.

Nature of Gullies and Ravination

The study of ravine has become a new aspect in the field of fluvial geomorphology. It creates much interest of study particularly in the European countries and United States of America. There are some pioneer soil scientists who have studied on ravine geomorphology viz. Brayan, K. (1941), Antev, E.V. (1952), Scheonwetter, J. (1962), Martin, P.S. (1963), Tuen , Yi Fu (1966) and Denevan, W.N. (1967), respectively.

In India, researchers like Tejwani, K.J. and Ahuja (1956), Gorries, R.M. (1957), Ahmed, E. (1968), Sharma, H.S. (1968) and Prasad, G. (1985, 87, 88) etc. have taken interest, taken an example from different parts of Paninsular India.

On the basis of the extensive field study and following the topographical sheets of the region, the present investigator has traced out the varying nature of gullies and further classified into four major ravine zones (Fig. 9.1 and Table 9.1).

Table 9.1: Nature of gullies and classification of ravine zones

S. No.	*Depth of gullies (m)*	*Nature of gullies*	*Major ravine zones*
1.	Below 3	Small gullies (G_1)	Zones of insignificant ravine of erosion
2.	3-9	Medium gullies (G_2)	Zones of moderate ravine erosion
3.	9-15	Deep gullies (G_3)	Zones of severe ravine erosion
4.	Above 15	Very deep gullies (G_4)	Zones of very severe ravine erosion

The Initiation and Development of Gullies

Slopes always have certain irregularities and depressions of varying size. Therefore, encountering such depressions separate thin streams merge together into more powerful ones. Such type of streams abruptly change the nature of slope, because of the greater kinetic energy and produce a noticeable gash which attracts the larger volume of water of atmosphere precipitation during the period of heavy rainfall and begins to grow in depth, breadth and upward direction along the slope regime. It is also observed that the gashes grow into very long and deep gullies beyond the valley slopes into the watershed spaces capturing ever new sections. Thus the progressive head of gullies constitutes a shear precipice (cliff) in their upper part which becomes a site of water falls during the rainy season. The water fall washes the participice and the gully grows upwards. In other words, the gully grows backward gaining more and more new sections of the inter-stream grounds. This process of gully growth is known as retrogressive of backward erosion. In view of the above mechanism four stages of gully developments have been recognised in the present study region.

I Stage: The first stage of gullies development is characterised by a rain rills or gashes having shallow depth of about 0.5 m in which streams of rain water concentrate. In this stage the longitudinal profiles of such terrain do not permit frequently the development of primary stage of cliffing with potholes. The above stage is known as the shallow hole stage of gully formation.

II Stage: The second stage of gullying begins with the formation of head precipice having approximate height ranging between 2 m and 15 m. The gully grows through the crumbling of the side walls at the head towards the watershed spaces. The view of the channel becomes very steep and uneven which reflects little relation to the profile of the slope. The mouth of the gully is separated by steep slope from the valley bottom into which it opens. It is termed as the tunnelling stage of gully formation. The right flank of Neolakhoh Pahari shows the tunnelling stage of gullies.

III Stage: The channel deepening is the common features of this stage. The mouth of gully generally reaches the level of the valley. The profile of valley floor becomes flatten and the gully grows wider and hill side waste accumulates in the lower part of the slope. The collapsing process dominates in this stage and thus it is termed as collapsing stage. (Agarhwa Pahar near Sheorampur) denotes this stage.

IV Stage: The fourth stage may be called as the stage of extinction of recession. Bed erosion diminishes and the head participate becomes flatten. The sides of the gully expose gentler slope covering vegetation and the gully floor accumulates erosional products. The area near Agarhwa Pahar and near Sitapur are the true indicator of the last stage of the gully formation.

Spatial Distribution of Ravine Erosion

On the basis of the nature and characteristics of gullies, the study region is classified into the following ravine zones (Fig. 9.1 and Table 9.1).

(a) Zones of Insignificant Ravine Erosion

This type of ravine zones is characterised by small gullies having depth of 3 m. These are generally seen in various patches along the right bank of Kewai basin and parallel to the lower reaches of Jhurai in the south-eastern part of the study region. Two small patches are noticed along the left flank of Jaiwanti nala. The northern and western flank of Bihara hill, the lower sections of Argarhwa pahar the toe slope of Khoh pahari, the sloping ground of Neolakhoh hills, the right flank of Yamuna river Rajapur, the lower sections of Paisuni, the side slope of Bharatpur hills, the area near Satna bus stand, Sitapur and the surrounding sections of a tributary stream of Paisuni joining at Raghav Prayag ghat from mouth towards source region clearly indicate the insignificant zones of ravine erosion. This type of ravine erosion can be easily seen in the entire region.

(b) Zones of Moderate Ravine Erosion

The vast area of the region falls under moderate ravine zones. The tributary streams of Paisuni, Ohan and Ganta (Kachchua, Samrani, Ganta, Kewai, Sukra and Geduwa) are highly ploughed by medium gullies. Kewai river catchment shows consistent gullying from source to mouth where about 960 hectares of land is washed away by ravination processes. Kewai produces maximum gullied land in the area. Ganta nala shows medium gullies at least in their upper reaches where it leaves the plateau region. The lower reaches of Kachchua nala is extensively ravaged where depth of gullying is reached between 3 m and 6 m. The middle and lower catchment area of Ohan and Samrani, the upper reaches of Barwa, the small tributaries of Banganga and Thothi Parari and Damodar nala are subjected as medium gullied land. Few interesting examples of medium gullied land are noticed from different localities surrounding the residual hills of the region. The side slope of Argarhwa pahar near Sheorampur about 10 km west of Karwi generally shows soil charged-water which generates medium to severe gullies having small boulders and pebbles over the thick blanketing of bare rock surfaces. The eastern flank of Sangrampur and Bihara hills are marked with gullies of about 5 m in depth. The sloping country denotes rills drawn parallel to each other from hill crest to bottom land. The panoramic view as taken from the top of Hanumandhara for the Chitrakut valley region, the surrounding alluvium sedimented area near Kamta Nath hills and the widely spaced gullied land near Sitapur are the part and parcel of moderate ravine zones. It is remarkable that these gullied lands are generally associated with finger tip tributaries over the thick alluvium cover. Most of the first order streams observed in this area initiate the formation and development of ravines. It is also apparent that the depth of ravines are closely concerned with alluvium cover and the parent materials seated in the lower section. In some localities kanker pan determines the depth of gullies. The topographical sheets are generally used for defining the ravine zones which are surveyed and published in 1974-75. The investigator has found that the depth of gullies is found to be more variable in the present time. One and half decades have been gone and the nature of the ravines are also found different in character at every place.

(c) Zones of Severe Ravine Erosion

The severe ravine zones are characterised by deep gullies with varying depth between 9 m and 15 m. The problem area can be marked in various scattered patches over the alluvium country mainly along the major streams like Paisuni, Banganga and Ohan. A continuous belt of severe ravines is found along the lower reaches of Paisuni near Kamta Nath hills, the surrounding area of Sitapur and along the Gara catchment area, a tributary of Banganga river.

Another patches of deep gullies (about 10 m in depth) are observed near Ainchwara village along the Ohan river where depth of alluvium deposits is noticed between 20 m and 40 m. A field survey was conducted along the river Paisuni between Sitapur and Karwi. The right flank of river Paisuni depicts severe and very severe ravinous land. The depth of gullies is found above 15 m along the river course but it is between 10 m and 15 m in depth in most of the surrounding areas.

(d) Zones of Very Severe Ravine Erosion

The very severe type of ravine erosion belt runs parallel to Yamuna river towards the north-east of the study region where very severe gullying (about 20 m in depth) is observed over the recent alluvium country. The thickness of alluvium deposits is measured as about 60 m (Fig. 9.2) below the ground level. The neighbouring area of Yamuna river is so seriously ploughed by ravines that there are only tiny remnants of artificially protected places of the ground over which are previously seated the villages of the region. The surroundings of Paisuni-Yamuna confluence indicates about 22 m of depth for gullies.

Few small patches of severe and very severe ravines are seen near Phatikshila, along the left flank of river Paisuni near Sirsaban and Jankikund and the surrounding area of the villages like Parati, Purwa and Delaura in the north of Sirsaban which are badly captured by very deep ravines. The area along the terrain of the tributary of Paisuni in the east of Kamta Nath hills clearly indicate very severe ravine zones. The aforesaid zones are more destitute bare of vegetation and possess thin grass lid and are found generally in V-shaped having vertical walls and flat bottoms. The zone represents a bewildering network of ravines and gullies.

Morphology of Ravines

The study of ravine, size shape, head characteristics, length and width and areal pattern are closely related to morphological study of ravines. Following to pioneer works of Tejwani, K.J. and Ahuja (1956), Gorrie, R.M. (1957), Sharma, H.S. (1968), Padmaja, G. (1976) and Prasad, G. (1985), the ravines of the region have been classified into four categories (Table 9.2) where depth of gullies has been considered for classification.

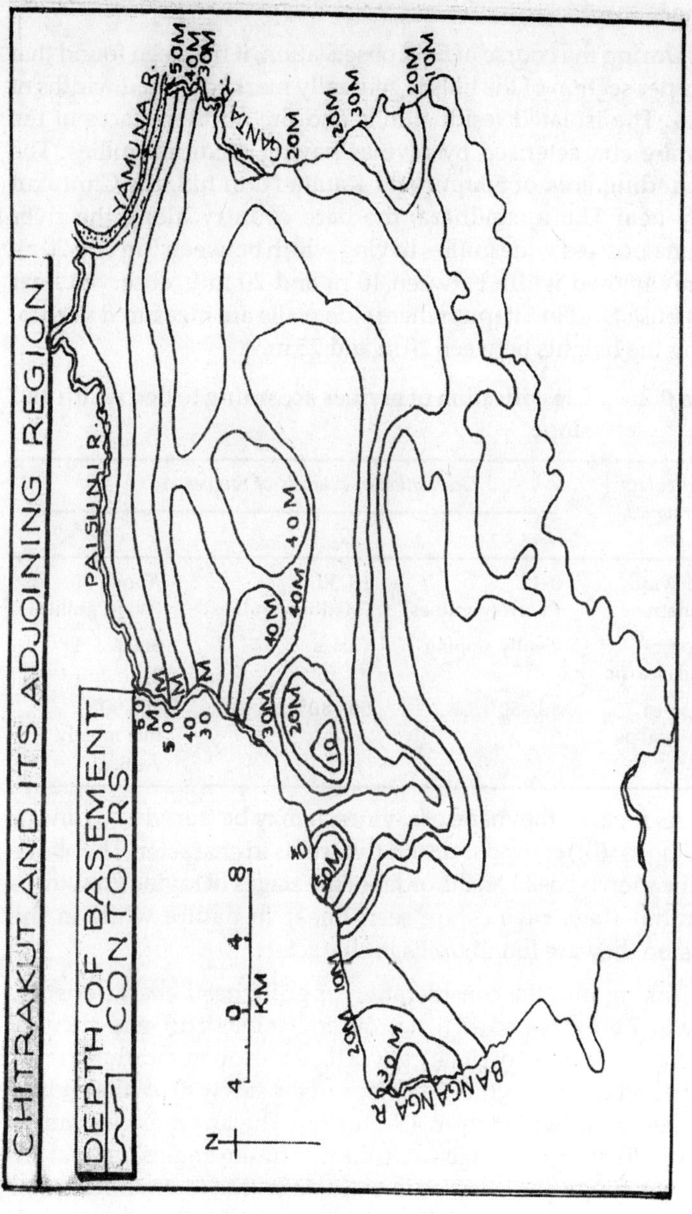
CHITRAKUT AND ITS ADJOINING REGION
DEPTH OF BASEMENT CONTOURS
YAMUNA R.
PAISUNI R.
GANTA R.
BANGANGA R.
KM
50M
40M
30M
20M
10M
40 M
N

Fig. 9.2

The bed width and slope of gullies differ from one place to another.

During the course of field observation, it has been found that the upper section of the hills is generally marked by bed widths of gullies. The isolated residual hills and the northern faces of the scarp are characterised by ravines having medium gullies. The surrounding area of Sitapur and Kamta Nath hill, the Chitrakut valley near Hanumandhara, the bare country along the river Yamuna possess wide gullies having width between 5 m and 20 m. The ravine bed width between 10 m and 20 m is observed near Satna bus stand in Sitapur where side walls are measured vertical having the heights between 20 m and 25 m.

Table 9.2: Classification of ravines according to bed width and slope

Particulars of ravines	*Description of symbols of ravines*		
	G_1	G_2	G_3
Bed width in metres	0-15 (Narrow gullies)	15-30 (Medium gullies)	Above 30 (Wide gullies)
Slope of head scarps	Gently sloping	varies	Steep
Slope of sub-scarps	40°-80°	50°-80°	50°-90° But mostly vertical

As regards the shape of ravines, it may be traced as (i) linear; (ii) balbous; (iii) compound; and (iv) trellis in character. The above classification is based on the order of the stages of ravine growth. In the initial stage ravines are seen linear in nature while in the recession they are found trellis in character.

Taking into the consideration of gully head characteristics, the ravines are grouped as: (i) narrow and pointed; (ii) semi-circular; (iii) notched; and (iv) digitate gully. These are found mostly inclined vertical and cave in section. The slope of the sidewalls of the ravines of the region differ between 45° and 90°. The lower angles can be observed in the initial stage while the maximum angles occur in the last stage. Slope angle varies in each stage due to environmental control.

Ravines are marked in V-shaped near Sheorampur, near Kamta Nath hills, Sitapur, along the left flank of river Yamuna, along the course of Paisuni near Raghav Prayag Ghat and near Pramod ban in the east of Kamta Nath hill respectively. It is because of commonly friable soil which is easily washed away by running water. It is apparent that the ravines maintain their V-shaped form in varying bed width. It must be seen the narrow long routes of ravines near the southern circle of Kamadgiri parvat. A narrow long passage is well defined in the ravaged land measured in the right flank of Paisuni.

Genesis of Ravines

The problem of ravine genesis is an acute in India and abroad. It needs scientific attention and enormous works have been done on the present awareness in the arid and semi-arid climates of the third world. But every attention of scientific in nature has been chocked out in Peninsular India. Various models have been proposed. The first model as propounded by Benett (1955) and Brice (1966) is the overgrazing model controlling for gully erosion.

The scientists like Brayan, K. (1941), Antev, E.V. (1952), Martin, P.S. (1963), Scheonwetter, J. (1962), Tuen, Y.F. (1966) and Denevan, W.M. (1967) of foreign countries and Ahmed, E. (1968), Sharma, H.S. (1968) and Dunn, J.A. (1938) of India advocated the concept of recent upliftment of ravine formation. Sharma attributes that the ravine erosion is closely related to the lowering of local base level of erosion initiated by the recent upliftment.

The present topographical behaviour is the outcome of accelerated stream action existed over the upland. The depth of Yamuna and Paisuni rivers indicate the history of past geological upliftment and thus the deepened channels must have accelerated the run-off and finally have generated rills and gullies along their courses where the lands were poorer in form in all geomorphic sequence. The present investigator advocates the upliftment theory of ravines genesis and prefers the impact of environmental factors such as geology, climate vegetation landuse pattern etc. If the upliftment is not possible in any case, the lithological characteristics and the sudden swift of climatic conditions will solve easily the present problem.

CONCLUSION

On the basis of the above discussion, it may be concluded that the area is drought prone where drinking water problem is most common in dry season. It is backward region and the associated terrain is highly ravaged by deep gullies. The soil erosion is the most serious problem of the region. All these problems have compelled the region behind the various serious problems relating agriculture settlement, communication etc.

10
Eco-Management and Planning

INTRODUCTION

Environment has been defined as the sum total of all conditions and influences that affect the development and life of organisms. The system preserves both physical and cultural attributes as a boon as well a curse for existing civilization. The so-called danger of environmental deterioration caused by the interactions between physico-cultural attributes has attained serious dimensions at places. Its study relates to environmental toxicology which in turn needs an urgent attention of eco-management in all possible techno-scientific ways.

Planning is considered to be a process of development through the exploitation and utilisation of all types of resources whether natural or human while environmental planning implies the optimal utilisation of the earth resources, both renewable and non-renewable for development activities conservation of what is rare and precious in nature and preservation of the quality of environment for the healthy growth of life (Singh, L.R. et al. 1983). Management implies a conscious choice from a variety of alternative proposals and further more that such a choice involves purposeful commitment to recognized and desired objectives. Environmental management is difficult to define because even the term environment in itself is complex as it is understood differently by different sections of society. The objectives of environmental management are complex, varied and even conflicting, and the alternative strategies are divergent (Singh, S. 1991). The concept of environmental management is generally related with the environmental model which assures the basic necessities with their important limits and includes the challenges to be faced and the policies to overcome the problem.

Human activities of modern 'economic and technological man' have distributed the harmonious relationships between the environment and man. Thus, environmental management is the process to improve the relationship between man and environment so that the quality of both, the environment and human society, may be improved. This improvement of relationships between man and environment may be achieved through check on destructive activities of man, conservation, protection, regulation and regeneration of nature. Environmental management, thus is related to the rational adjustment of man with nature involving judicious exploitation and utilization of natural resources without disturbing the ecological balance and ecosystem equilibrium (Singh, S. 1991). Environmental planning and management is therefore compromise between ecosystem and ecological balance and human material progress and thus environmental management must take into consideration the ecological principles and socio-economic needs of the society.

As a matter of fact, environmental management involves the protection of environment in all possible ways, enhancement of economic value of the environment and its resources and preservation of the environment for future generations. Environmental planning and management include the whole biological world. There is not any artificial boundary at regional, national or global level. But the processes of management and planning differ in different regions by changing human interests. The question of eco-management issues are indeed ever-widening. The purpose of eco-improvement is to develop it as a serious subject with a positive approach enabling us to wipe out the huge ecological deficit on the one side and prevent future environmental damage by making sustainable development on the other.

The aim of eco-management planning is to identify the problems of problem areas and the way of corrective actions which may lead sustainability in physio-cultural development.

Eco-management issues in India were raised by Pandit Jawahar Lal Nehru as early as 1957 when the question of environment in India was not even talked about. He wrote, "We have many large river valley projects which are carefully worked out by engineers. I wonder, however, how much thought is given before the project is launched to having an ecological survey of the area and to find out what the effect would be to the drainage system or to the flora and fauna of that area. It would be desirable to have such an ecological survey of these areas before the project is launched and thus avoid an imbalance of nature. There are so many organisations, programmes and projects which are actively engaged in the study of man-environment relationships, the effects emanating from such interactions and their possible remedial measures at international levels. It is a healthy sign that international cooperation is available for the mitigation of severe environmental problems affecting the mankind at global level.

INTRODUCTION TO MAJOR MORPHOLOGICAL PLANS ASSOCIATED WITH ENVIRONMENTAL CRISIS OF THE REGION

The Chitrakut and its adjoining region exhibit a variety of morphological plans ranging from the upland tops to the lowlands of Yamuna-Paisuni interfluves. The entire eco-system shows a vivid picture of terrain characteristics like recent flood plain along the river courses, piedment plain, plateau zone, flat topped hills, valley zones and scattered conical hills. The above morphological features constitute a complex outlook of lithologic sequence, pedological characteristics, relief irregularities, slope dynamics, dissection intensity, drainage frequency and texture variability from upland to fluvially dominated terrain. The entire region faces a lot of problems (Chapter 9) as a result of serious impact of various phenomenon upon the terrain and landscape of the region.

Environmental crisis is a common question appearing at every step, but it becomes more serious if it is growing by nature itself. The problems caused by ravination, water scarcity flood etc. are all caused by the existing natural phenomena. Thus, the conservation planning of natural resources and the improvement of geomorphic terrain is the first priority in maintaining the ecological balance of the region.

Man's interferences in natural phenomena have superimposed a new assemblage of terrain characteristics. He has used and reused the nature without much care for future generation and has changed the landscape according to his choice and needs. Obviously, the manmade crisis has precipitated more due to his involvement, positive or negative and unchecked interactions with nature. And thus the planning should be implemented at the level upon which man has engaged himself to consume the nature.

The nature becomes more poisonous if the physio-cultural eco-system together display the negative role. It is obvious that most of the serious eco-calamities are set in hazardous conditions due to ecological imbalance in nature. For example, the soil erosion is caused by the running water on the one hand and is further accelerated by faulty irrigational methods, deforestation, stone quarrying, unscientific, agricultural practices etc. on the other.

Physical environment has been dealt rather harshly by man through his advertent/inadvertent faulty actions. So, we must recognise the man, along with other life forms who have always been subjected to environmental stresses over which he has no control. But protective and evasive action is possible where advance sophisticated knowledge and methodology are available. We shall be foolish, if we don't take the opportunity to arm ourselves with information about these forms of environmental hazards. Where do such things occur? When can they be expected? Can warning system be made effective? What should be done when disaster threatens? Broadly speaking such information comes under the process of environmental protection. Although we have not given a full account of protection measures, we will explain how and where these dangerous natural phenomena operate the system and how they can encountered?

The nature and magnitude of eco-management planning have been suggested differently in various parts of the world depending upon the nature of the problems and the technologies available in the region. In the study region, the problems have been tackled in a superficial way because we do not have the requisite infrastructure to monitor the ecological changes and the possible hazards which threaten our life.

On the basis of the personal experience and some dialogue with the persons concerned in the environmental planning of the state, the author suggests various means and the measures which can solve the problems of environmental degradation in the study region. It consists of the following schemes:

- Afforestation plan and the conservation of biotic ecosystem;
- Management schemes of stone quarries;
- Soil moisture conservation and water holding plan;
- Ravination control techniques and other reclamation works;
- Watershed management;
- Groundwater management; and
- Tourism development.

Afforestation Plan and the Conservation of Biotic Ecosystem

Deforestation is the primary cause of land degradation in Chitrakut region. The uncontrolled and unwarranted deforestation have caused the greatest eco-disaster in the region. Although, the region is occupied by the dense mixed forest on plateau region, and the northern alluvial plain has only the scattered trees (Fig. 1.19). The forest cover consists mostly the natural flora and the types and the composition of plants are self nutrited in different localities. A systematic community of plants family cannot be recognised. The complexity and heterogeneity in plant community creates many problems of spatial arrangement and management.

The thick vegetal cover is most common along the water bodies and in highly moistured sloping areas occupied by the new sediments. Some examples of rich plant family are found along the Paisuni river course and along the debris slopes in different localities of Banda and Satna hills of the region. Thorny scrubs are seen in ravinous belt and along the isolated hills like Bihara, Bharatpur, Neolakhoh, Sheorampur, Satrajkifa, Gorra, Marpho, Chandra Maheshwari Matdar etc. Most of the residual hills show only grass cover resting over bare rock surfaces.

Deforestation is caused due to unchecked cutting of the trees by the contractors, overgrazing, stone excavation and morrum quarrying, supply of fuel, fodder and timbers for local people etc. The major part of Manipur hill forests and Satna forests in the south and south-east region have been deforested not only due to contractors encroachment for exploiting the timber wood but also due to overgrazing and stone excavation works. Overgrazing is also effective in Karwi and Sheorampur areas while the residual hills in the eastern, middle and south-western parts of the region are mostly devoid of vegetation due to stone and morrum excavation works. Illegal clearing of the forest by local people is also most common in thickly populated areas of Chitrakut Dham region. The tribal population generally supply the fuel wood to fulfill their primary needs to the local population of Sitapur and Karwi. Deforestation along the sloping ground of isolated hillocks in the eastern parts is dominated by the villagers impact i.e. clearing up the bushes and trees, digging up the soil, morrum and stony plates and open grazing works. This has led to the initiation of rills and gullies in the region.

Suggestions for Afforestation Plan

It is difficult to give general guidelines on the quantity and nature of deforestation which is still allowable in certain conditions, but a large scale plantation plan of significant plant species should be undertaken (Fig. 10.1) on the guidelines suggested below:

1. Most of the tribal populations are intimately tied to the forests as they depend for their entire needs upon the forest resources—fuel, fodder and small timbers. Thus, the plant species providing fuel, fodder and timber must be planted along the sloping and marginal land of the hills where the population of local people are most common. Patha population also demands the similar action.
2. The indigenous and exotic fast growing plant species are thought to be useful in plantation along the well drained sloping ground, ravination areas of the middle sections, barren lands between the agricultural fields and the lower section of the residual hills of the region. This type of production and protection forestry is seen along the residual hills where Eucalyptus plantation has been done.
3. The agricultural areas of plateau top and the faces of the scarps should be harvested in a very careful and controlled manner so as to avoid soil erosion and tree damage. The plantation of rooting plants (*Mangifera indica, Madhuca indica, Dalbergia sisso, Aratchdica indica etc.*) are the best solutions in stony bed areas.

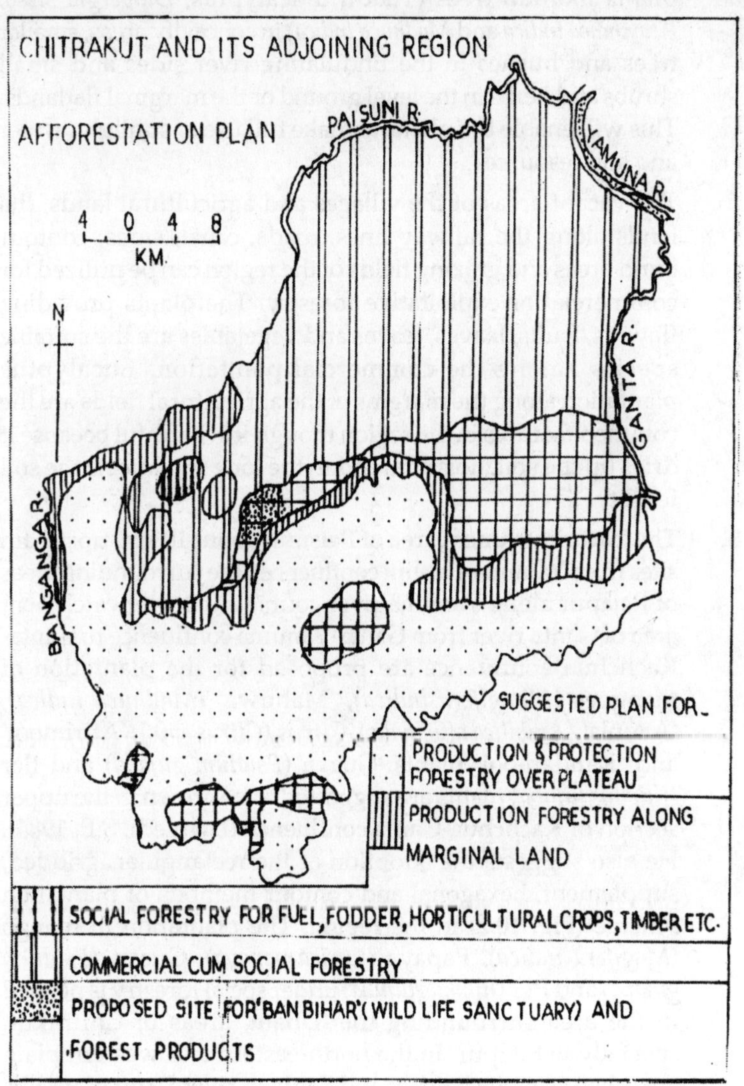
CHITRAKUT AND ITS ADJOINING REGION
AFFORESTATION PLAN
PAISUNI R.
YAMUNA R.
4 0 4 8
KM.
N
GANTA R.
BANGANGA R.
SUGGESTED PLAN FOR_
PRODUCTION & PROTECTION FORESTRY OVER PLATEAU
PRODUCTION FORESTRY ALONG MARGINAL LAND
SOCIAL FORESTRY FOR FUEL, FODDER, HORTICULTURAL CROPS, TIMBER ETC.
COMMERCIAL CUM SOCIAL FORESTRY
PROPOSED SITE FOR 'BAN BIHAR' (WILD LIFE SANCTUARY) AND FOREST PRODUCTS

Fig. 10.1

4. The areas of the concave hill slopes, river sides and undulating country must be managed by planting suitable plants like tall trees (Tendu, Eucalyptus, *Dalbergia sisso, Aratchdica indica* and *Madhuca indica*) in concavity areas, smaller trees and bushes in the undulating river sides and small shrubs and herbs in the level ground or the marginal flatlands. This will enable the plants to make full use of sunlight, water and soil resources.
5. The vacant areas of the villages and agricultural lands, the lands along the railway lines, roads, canal, sides, contour bund areas and grazing fields of the region can be utilized for commercial or exploitative forestry. The plants providing flowers, fruits, leaves, leaves and vegetables are the suitable species among the commercial plantation. Eucalyptus plantation along the margins of the agricultural fields are the common features in the region though it is harmful because it dries up the soil moisture, and in the long run affects the soil fertility.
6. The lower catchment area of Paisuni (about 10 km. upstream sites from Yamuna-Paisuni confluence), the surrounding areas of Rajapur along Yamuna river course and lower catchment area of Ganta river from Ganta-Yamuna confluence to Ganta-Kachchua confluence are proposed for the plantation of mangoes (*Mangifera indica*), 'Mahuwa' (*Madhuca indica*), 'Aaunla' (*Ambilica officinalis*), Citrus (*Citrus spp.*), 'Attrimool' and 'Kandamool' plants. Guava (*Psidium gujava*) and Ber (*Jijiphus jujuba*) plants are suggested to be grown in the upper section of Kachchua-Ganta confluence (Dwivedi, S.P. 1984). He also suggests the adoption of the rectangular, gridded, supplement, hexagonal and contour methods of plantation scheme (Fig. 10.2) in the region. The plantation of mango (*Magifera indica*), Papaya (*Carica papaya*), Guava (*Psidium gujava*) and Ber (*Jijiphus jujuba*) under social forestry is needed in the area surrounding the 'Dham' areas of Chitrakut, specially in Sitapur. In the north-eastern and western plain areas between the Paisuni and Baghain rivers fruit plant like Papaya (*Carica papaya*) can be grown for maximum economic gains and also for the preservation of the floral ecosystem.

7. Fuel, fodder and timber plants should be grown in the ravinous belt in the south of Sitapur. The area along Satna bus stand and the north-west sites of Kamtanath temple should be managed by fire wood plantation. Chaturvedi, A.N. (1985) has also proposed similar measures to utilizes the degraded land by planting firewood because the land once put under tree forming will improve slowly and surely the ecosystem.
8. It would be better if the area in the south of Sitapur be converted into 'Ban Bihar' a National park, the botanical garden and a research station. The forest department should take up the job and establish a research centre to conserve the flora and fauna and also to provide attraction to the tourists. The forest division, Manikpur Banda has introduced 113 species of trees, 36 species of herbs, 18 species of creeping plants and 20 species of grasses to cover the entire area. The proposed research centre will experiment and suggest precisely the nature of plants to be grown at the specific spots.

In addition to above mentioned afforestation schemes, the conservation of biotic ecosystem including the preservation of plant species, wildlife and living resources available for the tribal population is of vital importance. The large scale production of wild citrus (*Citrus spp.*), Tendu leaves (from about 1200000 trees) and Karaunda fruits (Caraunda carandas) in Manikpur areas, tree roots in Sheorampur Chitrakut areas and the common bushes and trees (locally termed as Churk, Simali, Kataiya and Thopa) providing leaves as a food resources must be conserved. Maximum efforts must be made to conserve the fauna. The fisheries department has prepared a plan for the growth of fisheries including preservation of specific fish varieties. Accordingly, the large amount of 'Rohu fish' of Paisuni river (generally seen near Sati Anusuiya Ashram and Sphatikshila) should be preserved and nurtured in various ponds and reservoires of the region.

Legislation for the Protection of Biotic Environment

With respect to environmental problems, various legal remedial actions have been taken by the State Governments. For the safeguarding of forest ecosystem, commercial exploitation is totally prohibited specially in the degraded areas of the region. The protection of reserved and protected forests from fuel starved

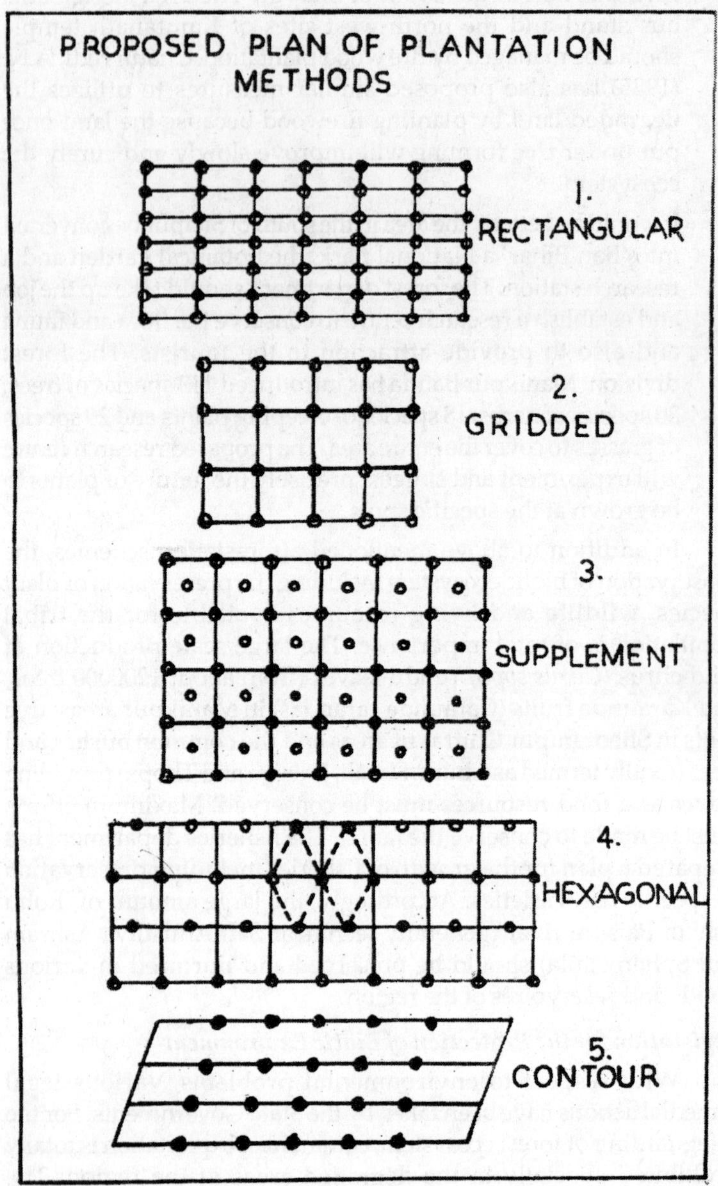
PROPOSED PLAN OF PLANTATION
METHODS
1.
RECTANGULAR
2.
GRIDDED
3.
SUPPLEMENT
4.
HEXAGONAL
5.
CONTOUR

Fig. 10.2

villagers and fodder starved cattles have been taken up on priority basis. Obviously, for this public support can be elecited only when there is a meaningful development programmes for supply of goods and services by the State Governments to meet the villagers genuine needs, the survival of Kols and to raise their standard of living (Srivastava, R.C. and Prakash, P. 1991).

The implementation of first 'Forest Working Plan' in 1896 prepared by Blanchfied for the whole Bundelkhand region indicates that the forest of the area was being exploited without practising forest management techniques. The Kols were not given concessions or rights to gather minor forest produce whereas the landlords enjoyed the facility. The right to collect Mahuwa flower, wood for building construction and manufacturing of agricultural implements and grazing of cattle was given to the landlords as per the notification of 11 May, 1892 by the Forest Department. In the first governments policy on forests (1894) certain restrictions were imposed for the use of forests and forests produce which affected mostly the tribals. In 1952 the policy of 1892 was reassessed and even more restrictions were imposed on tribals which created a tension between the many tribal communities and the Government.

In 1894 they became 'rights and privileges', in 1952 they became 'rights and concessions' and now they are being regarded as 'concessions'. A nationwide uproar was marked in 1982 against the forest bill amending the Forest Act of 1927 which gave excessively authority to the forest officials to control the entry of poor tribal people who need forest produce for their survival. The Committee for Review of Rights and concessions in the forest recommends that "exercise of rights and concessions" should be restricted for the people residing at a maximum distance of 8 km from the existing forest. As regards the collection of firewood of headloader, the entry and collection are restricted and the sale of headloads in urban areas is planned to be completely banned.

The new Forest Policy and the Forest Conservation Act December 1988 have many positive points but do not look positive for forest dwellers. Roy Burman Committee on Forest Policy and Tribal Development points out the benefits of tribal people by imaginative forestry programmes and conservation and reorganisation of their traditional skill.

Bundelkhand Vikas Nigam has launched the Tendu operation scheme in May 1982 only for 8 units of Mainkpur forest range. Still the Tendu leaf contractors exploit the tribals with the connivance officials of Van Nigam. Bundelkhand Vikas Nigam must weaken the contractors and cover the rest units of all the forest ranges. The ownership rights over the forest produce to the tribals should be controlled under 'Sanat system' and the additional employment opportunities must be provided by creation of household industries on the forest products in the region.

Management Scheme of Stone Quarries

A large scale disaster of land degradation is noticed throughout the region where the stone excavation is common (Fig. 10.3). The stone excavation has initiated the mass movement and slumping of rocks. The problem in Manikpur area and around Karwi has become more serious by large scale excavations by unauthorised contractors. Stringent actions must be taken against the offenders to curb their illegal actions. Wherever the quarrying poses serious threats the contour bunding of varying heights must be erected and the land should be covered with new plant species. The quarrying of morrum is a common features along the residual isolated hills (Fig. 10.3). The morrum and sand accumulation are seen continuously in association with agricultural fields. In Bharatpur and Sheorampur areas the problem has reached serious dimensions. Limestone quarrying is generally observed near Gupt-Godavari and Khoh areas. The areas of morrum quarrying should be provided with counter bunds. The wire nets filled with rocks having concrete base should be provided at suitable places. The areas of non-perennial fingertip tributaries should be treated by such constructions. It is generally observed that the isolated hills located in the north of the scarplines are mostly devoid of vegetation. Only thorny scrubs appear in rainy months. Thus, the barren field produces the larger quantity of morrum, bajari, pebbles, boulders and sandstone plates on the stony beds. The rounded granitic knobs are also observed. The circles marked in the map (Fig. 10.3) around the hills, demarcate the perimeters for contour walling of the problem areas.

Soil Moisture Conservation and Water Holding Plans

The proposal of soil moisture conservation and water holding plans has vital importance because the region is a water scarcity zone and is covered under Drought Prone Area Programme (DPAP) of the state. The study conducted by Brockman, L.D. (1929), the Regional Soil Laboratory, Jhansi and the Soil Survey Organisation, U.P. (Mehrotra, C.L. and Gangwar, B.R.) reveal the facts that the soils of the region contain very low to moderate soil moisture intensity (Fig. 10.4). The soil of upland areas (Patha soil and popularly known as Ranker); the coarse grained brown soil (Parwa), the shallow black soil and the riverine soil (Tari, Kachchar and Rakar) have a very low, low to slightly moderate, uniform and average and moderate soil moistures respectively. The compact nature of sandstone does not allow the rainwater to percolate and go deeper, while the limestone areas are pervious and the water percolates freely. The granitic covers having rocky knobs along the residual hills and the sloping ground of isolated hills having bare surfaces clearly indicate the poor soil moisture content. On the contrary, the rectilinear slope profiles covered with thick vegetal properties show sufficient soil moisture.

The Ground Water Investigation Division, Lucknow has conducted the geophysical survey of the region and suggested the construction of contour bunds to conserve and improve the soil moisture and water holding capacity of the soil (Fig. 10.5). Under the Drought Prone Area Programme (DPAP), the water scarcity region of the north Yamuna plain is contour bunded. These bunds are 3 m to 6 m in height and need further extension in length and height. Plantation of new plant species and shrubs, construction of dams and reservoirs (Fig. 10.6) and further extension of agricultural and irrigational plans may be suitable for water holding capacity of the soil. The plan of graded bunding must be applied because the previous experiments have clearly indicated that the level bundings were found unsuitable in different states of India. The dams and reservoirs should be constructed in the lower sections of the river channels.

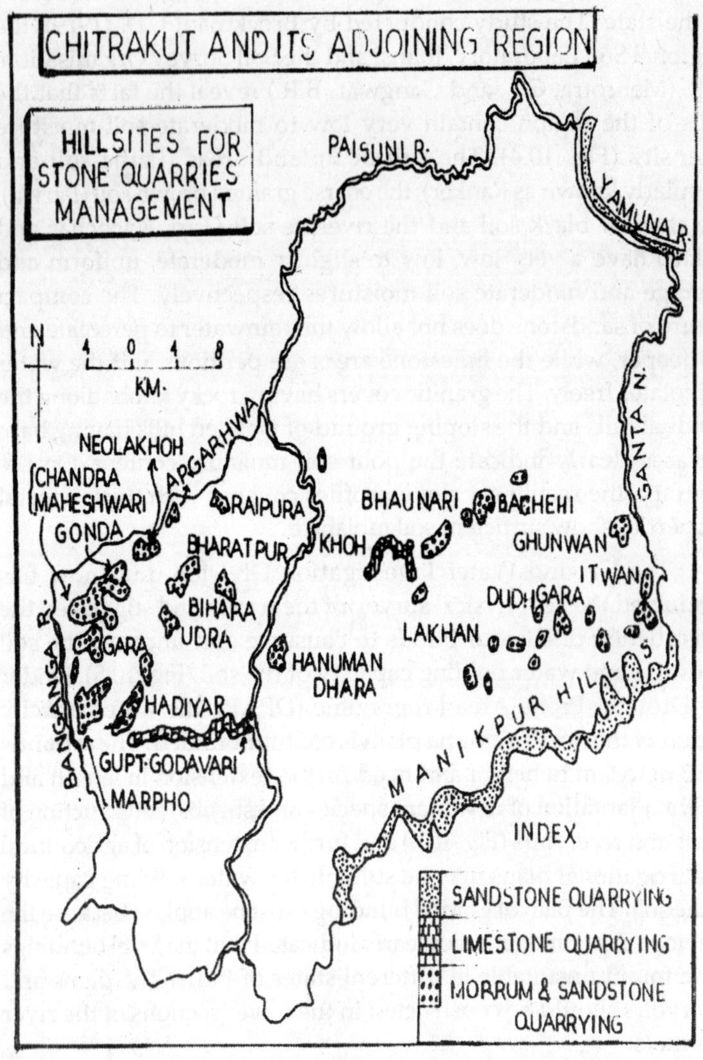

CHITRAKUT AND ITS ADJOINING REGION
HILLSITES FOR STONE QUARRIES MANAGEMENT
PAISUNI R.
YAMUNA R.
GANTA N.
N
KM.
NEOLAKHOH
CHANDRA
MAHESHWARI
GONDA
ARGARHWA
RAIPURA
BHARATPUR
KHOH
BHAUNARI
BAGHEHI
GHUNWAN
ITWAN
DUDH GARA
LAKHAN
BIHARA
UDRA
GARA
HANUMAN DHARA
HADIYAR
GUPT-GODAVARI
MARPHO
BANGANGA R.
MANIKPUR HILLS
INDEX
SANDSTONE QUARRYING
LIMESTONE QUARRYING
MORRUM & SANDSTONE QUARRYING

Fig. 10.3

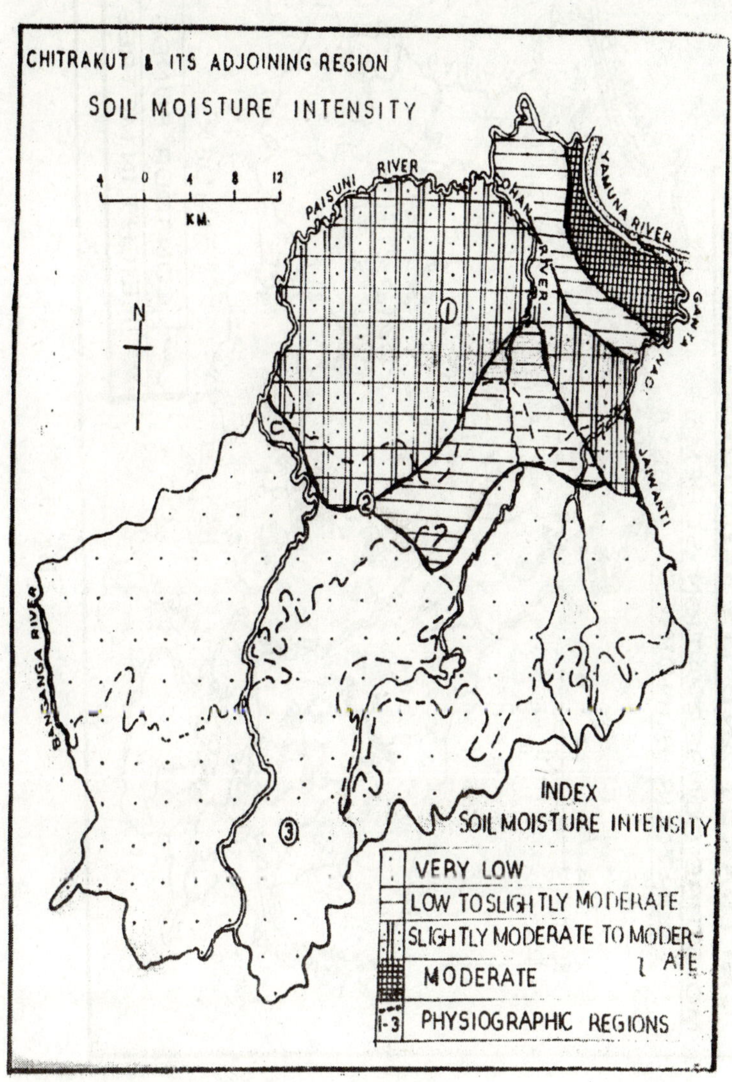
CHITRAKUT & ITS ADJOINING REGION
SOIL MOISTURE INTENSITY
4 0 4 8 12
KM
N
PAISUNI RIVER
OHAN RIVER
YAMUNA RIVER
GANTA NAG
JAIWANTI
BANGANGA RIVER
1
2
3
INDEX
SOIL MOISTURE INTENSITY
VERY LOW
LOW TO SLIGHTLY MODERATE
SLIGHTLY MODERATE TO MODERATE
MODERATE
1-3 PHYSIOGRAPHIC REGIONS

Fig 10.4

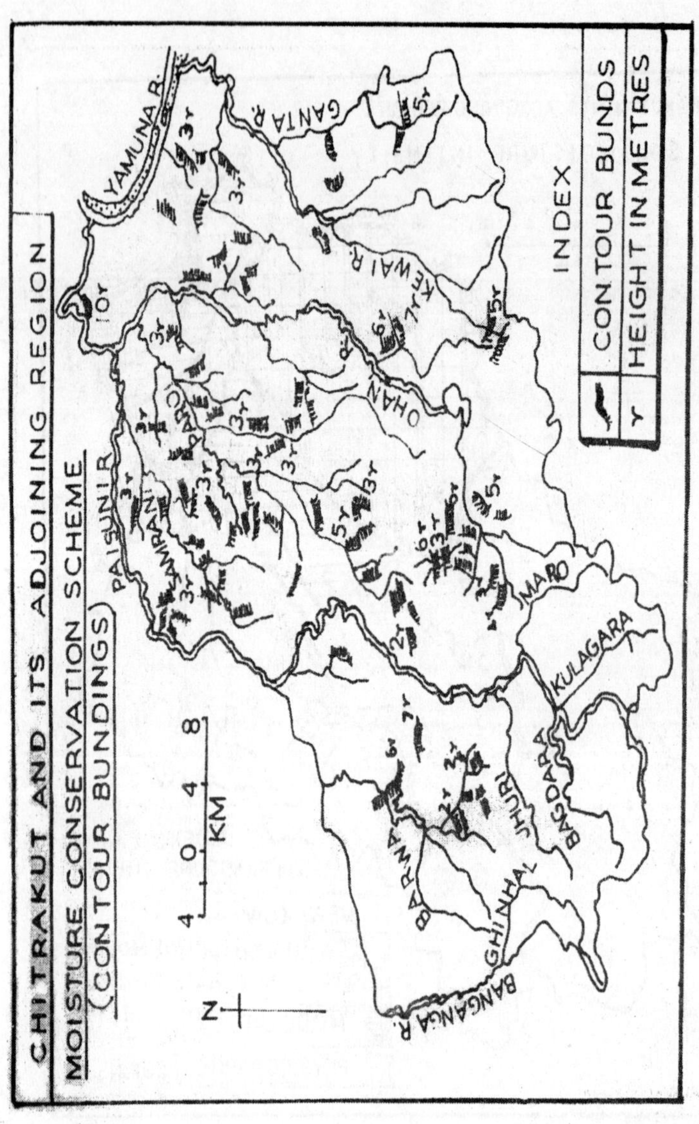
CHITRAKUT AND ITS ADJOINING REGION
MOISTURE CONSERVATION SCHEME
(CONTOUR BUNDINGS)
N
4 0 4 8
KM
YAMUNA R.
PAISUNI R.
GANTA R.
KEWAI R.
OHAN R.
MARO
KULAGARA
BAGDARA
JHURI
BANGANGA R.
INDEX
CONTOUR BUNDS
HEIGHT IN METRES

Fig. 10.5

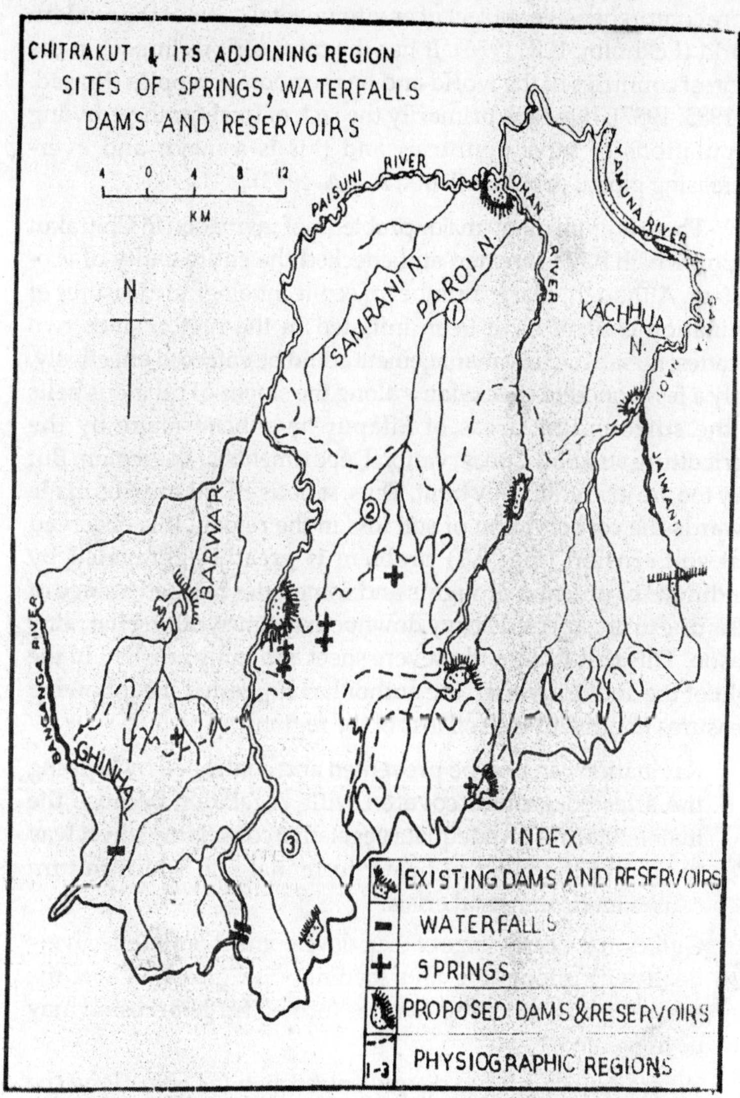
CHITRAKUT & ITS ADJOINING REGION
SITES OF SPRINGS, WATERFALLS
DAMS AND RESERVOIRS
4 0 4 8 12
KM
N
PAISUNI RIVER
OHAN RIVER
YAMUNA RIVER
SAMRANI N.
PAROI N.
KACHHUA N.
BARWAR
JAIWANTI
BANGANGA RIVER
GHINHAI
INDEX
EXISTING DAMS AND RESERVOIRS
WATERFALLS
SPRINGS
PROPOSED DAMS & RESERVOIRS
PHYSIOGRAPHIC REGIONS
1-3

Fig. 10.6

Ravination Control Techniques and Other Reclamation Works

Land degradation which is expressed as soil erosion or ravination is a well known phenomenon and has become a much more comprehensive subject of environmental crisis of the modern world (Eckholm, E.P. 1976). It has the most serious impact in the poorest countries of the world and on the poorest peoples (Prasad, G. 1985, 1987). It means primarily the lack of food for the growing populations of most countries and this is a major and ever-increasing global problem (Vink, A.P.A. 1983).

There are still many major problems of ravination in Chitrakut region which have threaten and checked the entire entity of eco-system. Although, a large number of eco-technological measures of ravination control have been initiated in the region since two decades; no satisfactory management could be enforced effectively. Only a few concrete check dams along the slopes of ravinous belts in the surrounding areas of Sitapur have been made by the Agriculture and Soil Conservation Departments of the region. But they too are inscientifically built. Thus, serious efforts must be made towards the conservation of soil loss in the region. It is observed that soil erosion (Fig. 9.1) problem is greatly aggravated by conditions of unusual droughts and sometimes reached a stage of crisis due to unexpected heavy downpour at times, during the rainy season. This results into the severe sheet and gully erosion. In the light of the above disaster, the author has suggested the following measures to prevent the soil loss of the region:

1. Ravination can best be prevented and controlled by keeping the affected ground covered with vegetation because the maintenance of an adequate vegetative cover is the surest way to control the burst of heavy torrential rain which in turn causes environmental crisis.
2. Gullies in the early stage of their development can be effectively dealt with ploughing and seeding with grasses. Once the grasses grow and a sod cover is formed further erosion may be prevented.
3. Where gullies are too deeply entrenched (as seen along the Paisuni river course), and it is difficult to plough the land, it is necessary to construct retaining walls perhaps of brush wood

or wire netting with a straw gaurd to catch and hold the eroded soil. Great chasms may demand the construction of concrete or stone dams.

4. Terracing is important in sloping land under certain conditions. But, if used improperly, it can be ineffective and may be harmful.

5. The sloping country which is prone to erosion and cultivated should be practised by contour ploughing because ploughing up down slope will clearly assist runoff and predispose towards gully formation.

6. Strip cropping is a suitable measures in most of the cultivated areas. Strips of grass and soil binding leguminour crops between annually planted cereals should be managed in cultivated land to check soil loss.

7. The large cultivated fields should be divided into the smaller fields surrounded by walls or hedges.

8. The steep slope areas of the upland should be afforested by new plant species because the foliage protects the ground and the roots help to hold the soil in place.

9. The limited grazing may be allowed in the problem areas. The animal numbers must be reduced.

10. The cropland which is unsuited to crop growing must be changed into pasture land.

11. Crop cover and crop rotation should be maintained which will give seasonal protection to the soil.

12. The severely eroded and uneconomically redeemable land must be retired for the permanent protection of grass or trees.

On the basis of the aforesaid general guidelines of preventing the soil by ravination hazards, the present investigator has selected the severely ravaged areas from different localities of the region and suggested the suitable ravine management plans keeping in view of terrain characteristics, pedological balance, slope dynamics and drainage orientation existing at specific places. The suggested ravine reclamation plans are as follows:

1. The ravines themselves can be brought under cultivation in Sitapur areas where these are sufficiently wide by the

formation of a series of terraces one upon another along the slope. Where ravines are moderately wide they can be terraced and must be used for growing horticultural crops like grafted mangoes, guavas etc. The ravinous land surrounding the Jankikund and Pramod Ban situated along the Paisuni river and around the dissected isolated hills near Sheorampur need the plantation of horticultural crops to minimize the serious impact.

2. The serious regime of ravinous belt of the region should be managed by ponding schemes associated with new plantation. The construction of small tanks will fulfill the irrigational demands in the neighbouring areas. It is notable that the area around the Kamta Nath temple and Lakshman Pahari has needed the construction of a large size tanks where the tourists can take bath before the Parikrama of Kamadgiri. The ponding condition must be surrounded by trees.
3. The small and medium gullied lands which are located nearby the small towns like Pahari, Rajapur, Sheorampur, Bharatkup etc. should be used for settling landless labourers, lower class colonies and as well as for playground.
4. The steep gullies and very severe ravine zones in the north-east of Raghav Prayag ghat may be reclaimed by retiring them for fuel forests.
5. The medium gullied land specially in Chitrakut Dham area should be converted into 'Ban Bihar' associated with newly attractive plants where tourists can take rest and enjoy with the sweet vendure of environment. This will provide more attraction to tourists or pilgrims (Darshanarthi) who migrate to obtain peace at this holy centre.
6. The control of ravines in flat land area of Yamuna-Paisuni plain in the north of the region can be made by construction of marginal bunds, by diversion channels in the catchment feeding ravines, by the construction of check dams or gully plugs in the ravines at proper intervals, by toning down or easing the steep slopes on the gully banks to the angle of natural repose and sodding the slopes with vines and grasses having soil-binding tendencies, by planting suitable fast growing trees and shrubs along the banks of the major rivers

(like Paisuni, Ohon and Ganta) and in the cut up areas which are unfit for the purposes by the construction of wire crate walls inside the 'Khad' and by the introduction of better farming practices and soil conservation measures in the tablelands of the catchment (Ray Chaudhari, S.P. 1966).

7. The tourists sites of Mandakini (Paisuni) valley wherever ravaged by ravines or affected by growing gullies in Chitrakut holy centre must be converted into "Pakka Ghat."
8. Land ownership should be planned in ravinous land. It is apparent that the gullied lands are under the ownership of three different agencies, i.e. Government, Goan Samaj and private ownership. Patches of Goan Samaj land should be eliminated from reclamation blocks by exchange of land. Generally land consolidation work should also be done. Ownership boundaries should be allowed in the way of scientific planning (Sharma, H.S. 1979).
9. The irrigation plan should be prepared for the crisis land of the region and holy if irrigation is possible, investigations be further carried out for creation of reclaimed blocks for agriculture.
10. Crop management practices including the use of a legume or grass crop in rotation should be adopted in an integrated watershed planning of ravine lands. Cover crop should be harvested when the main crop is harvested. Continuous farming by planting rows of crops and grasses and terracing may be followed in sloping areas.

Watershed Management

The problem of eco-management planning can be better understood with reference to watershed areas. And obviously the need for watershed management is vital for an efficient management of the ecosystem in the region. It will include both the perennial as well as non-perennial segments of the river channels and landlocked water bodies (i.e. reservoirs, canals, tanks and ponds either natural or artificial in nature) of fluvially dominated terrain. The planning for the non-perennial segments has already been discussed in the earlier sections. The perennial segments include the eco-management plans regarding the river regimes and landlocked water bodies wherein the improvement of river catchment and management of waterfalls, rapids, reservoirs and ponds have their greater importance in the present attempt. The plan also includes the assessment of navigational sites in major river channels for the welfare of local people and also in view of developing beauty spots from the tourism point of view. The remaining aspects of watershed management for example ground water management (10.2.6), tourism development (10.2.8) and ravine control and reclamation works (10.2.4) have been discussed in detail separately in subsequent sections. The maps show the different sites along the fluvial zones of the region which need the proper watershed management schemes. The prominent among the schemes are as follows:

1. The major river basins like Paisuni, Ohan, Ganta and Banganga need the large scale management in their catchment areas. The tributaries of the major channels have their uncertain flow paths and the catchment areas are yet to be delimited. The water divides are found zigzag and strech in a haphazard way. Therefore, the identification of interfluves and water divides (Fig. 1.15) must be considered at first priority. The Ground Water Investigation Organisation Drought Prone Area Programme of State Government has identified the fluvial zones of the rivers to control the soil loss in the region.
2. The meandering flow path of river Paisuni (e.g. in the north-west of Chitrakut Dham and before and confluence site with Ohan and Yamuna river in the north) and Ohan (near the confluence site with Kachchua nala) must be converted as a straight fluves otherwise it may produce ox-bow lakes as already exists in the north of Paisuni river near Yamuna-Paisuni confluence (Fig. 10.8).
3. The encroachment by the construction of pakka ghats and temples along the river Paisuni in Dham areas have shortened the width of the river basin which in turn causes the flood problems. The unplanned constructions projected towards the valley must be prohibited and further extension should be made parallel to the river course in the upper sections of the valley.

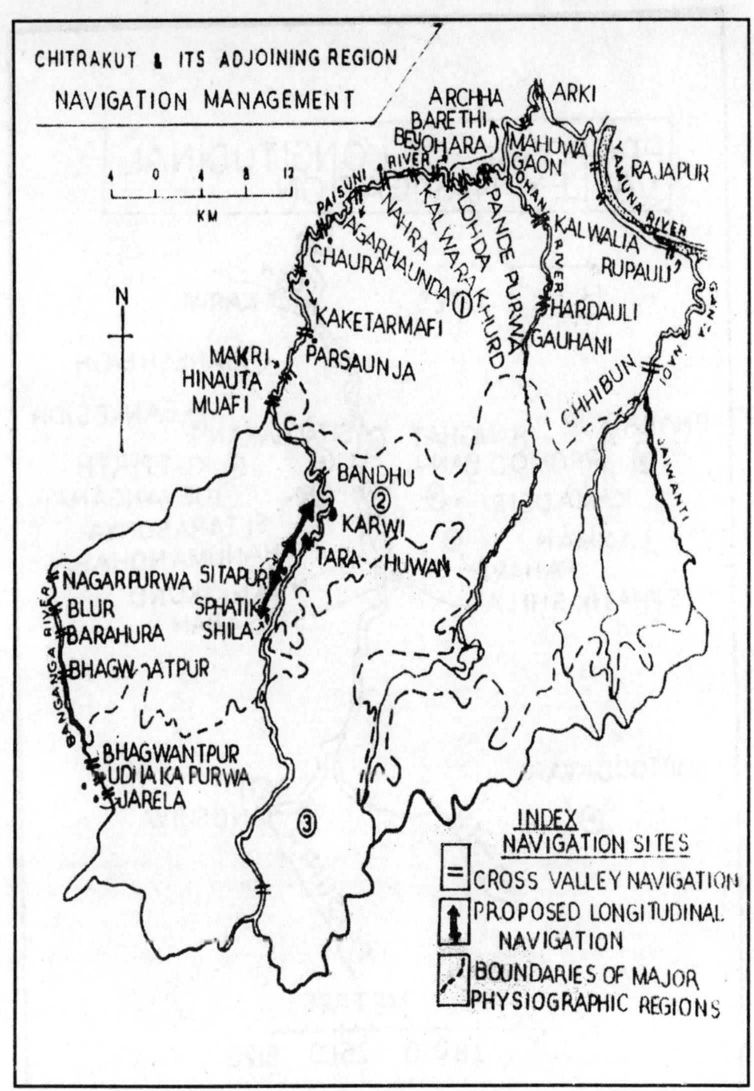
CHITRAKUT & ITS ADJOINING REGION
NAVIGATION MANAGEMENT
4 0 4 8 12
KM
N
ARCHHA
BARETHI
BEJOHARA
ARKI
MAHUWA GAON
RAJAPUR
YAMUNA RIVER
PAISUNI RIVER
NAHRA
KALWARA KHURD
LOHDA
PANDE PURWA
AGARHAUNDA
CHAURA
OHAN RIVER
KALWALIA
RUPAULI
HARDAULI
GAUHANI
KAKETARMAFI
MAKRI,
HINAUTA
MUAFI
PARSAUNJA
CHHIBUN
JAIWANTI
BANDHU
KARWI
TARA-
HUWAN
SITAPUR
SPHATIK
SHILA
NAGARPURWA
BLUR
BARAHURA
BHAGWANTPUR
BANGANGA RIVER
BHAGWANTPUR
UDHA KA PURWA
JARELA
INDEX
NAVIGATION SITES
CROSS VALLEY NAVIGATION
PROPOSED LONGITUDINAL NAVIGATION
BOUNDARIES OF MAJOR PHYSIOGRAPHIC REGIONS

Fig. 10.7A

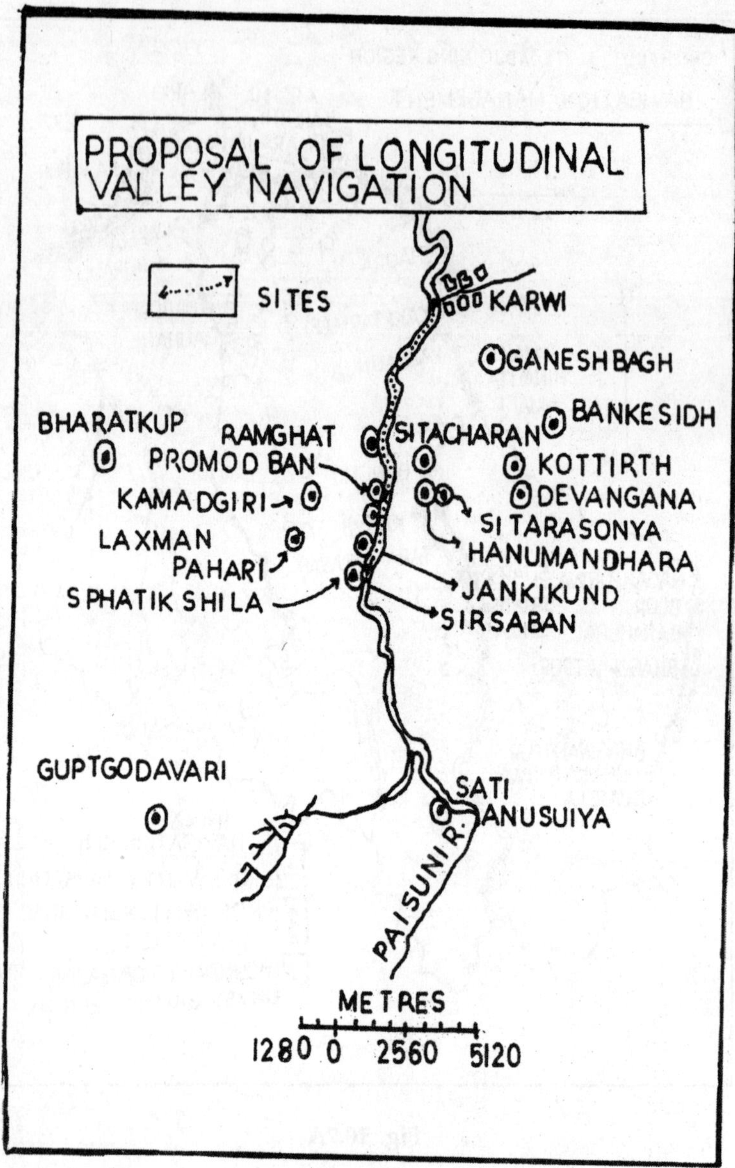
PROPOSAL OF LONGITUDINAL VALLEY NAVIGATION
SITES
KARWI
GANESHBAGH
BANKESIDH
BHARATKUP
RAMGHAT
SITACHARAN
PROMOD BAN
KOTTIRTH
KAMADGIRI
DEVANGANA
SITARASONYA
LAXMAN
HANUMANDHARA
PAHARI
JANKIKUND
SPHATIK SHILA
SIRSABAN
GUPTGODAVARI
SATI
ANUSUIYA
PAISUNI R.
METRES
1280 0 2560 5120

Fig. 10.7B

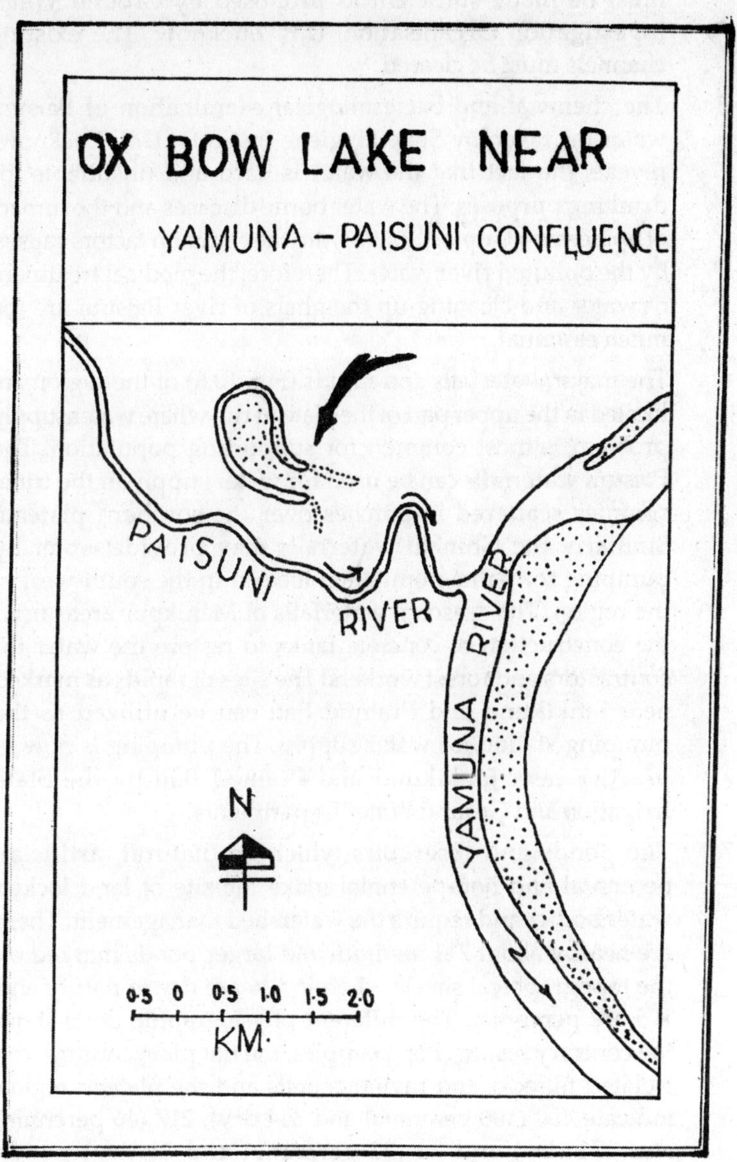
OX BOW LAKE NEAR
YAMUNA – PAISUNI CONFLUENCE
PAISUNI
RIVER
YAMUNA
RIVER
N
0·5 0 0·5 1·0 1·5 2·0
KM·

Fig. 10.8

4. The quantity of surface water discharge in major channels must be made sufficient as proposed by Ground Water Investigation Organisation, U.P. Lucknow. The existing channels must be cleared.

5. The chemical and bacteriological examination of Paisuni water (as taken by State Hygiene Institute, U.P. Lucknow) reveals the fact that the water is hard and unsuitable for drinking purposes. The water borne diseases and the impact of epidemics (Gupta, J.P. 1981) are the common factors caused by the polluted river water. Therefore, the medical treatment of water and cleaning up the ghats of river Paisuni are too much essential.

6. The major waterfalls and rapids (Fig. 10.6) of the region are located in the upper part of the plateau rim where water supply problem is most common for supporting population. The Paisuni waterfalls can be used for water supply in the tribal colonies scattered in patches over the southern plateau. Similarly the Ghinhal waterfalls may provide water by pumping for patha population located in the south-west of the region. The seasonal waterfalls of Manikpur areas need the construction of concrete tanks to restore the water for contractors and forest workers. The sites of rapids as marked near Jankikund and Pramod Ban can be utilized as the pumping station for water supply. The pumping is now in practice near Jankikund and Pramod Ban by the State Irrigation and Ground Water Departments.

7. The ponds and reservoirs which are natural, artificial, perennial and non-perennial make the site of land locked water bodies and require the watershed management. There are nearby about 711 medium and larger ponds marked on the topographical sheets wherein 536 are dry in nature and 175 are perennial. The different physiographic units show the contrary results. For examples, the flat plain country, the isolated hillocks and ravinous belts and the plateau region indicate 450 (106 perennial and 344 dry), 217 (46 perennial and 171 dry) and 44 (23 perennial and 21 dry) ponds respectively. Most of the ponds have been constructed in the form of relieve works during Chandel, Bundela, the British

regime and a few after independence. Tanks have also excavated either by villagers in cooperation with or by Jamindar or Talukedar of the area to meet the various requirements such as drinking purpose for cattle and other domestic uses. The tanks over the top of the plateau in the tribal localities génerally dry up after monsoon period due to underlying limestone strata which percolates the rain water suddenly just after the rain. The bottom of these tanks must be treated suitably otherwise lack of water may be found. A high contour wall shaded with suitable plant species and shrubs is also needed around the ponds. They should be enlarged in depth, width, length and perimeters and can be developed for fisheries.

In the pre-independence period, some reservoirs wets constructed by the Jamindars in the region. The existing dams and reservoirs (Fig. 10.6) like Paisuni, Barwa, Ohan (storage capacity as 1069630 KLD), Jaiwanti, Barar and Donamaphi generally supply water by pump canals. These reservoirs mostly suffer from seepage losses owing to jointed nature of Kaimur sandstone on which these have been constructed. In the wet monsoon season these reservoirs contain sufficient water and can be utilized for drinking purposes after being treated properly. The four sites (e.g. near Sitapur along the Paisuni river at the confluence of Paisuni and Ohan, in the middle course of Ohan and at the confluence site of Jaiwanti nala with Ganta river) for dams and reservoirs are proposed (Fig. 10.6) in view of remedial actions in the water scarcity zones of the region.

8. The development of navigational facilities are the major objectives of watershed management in the region. The navigation will provide the transportational facilities as well as increase the recreational interests from tourism point of view. Although the cross navigation (Ferry points) is common in Yamuna, Paisuni and other perennial streams of the region for local peoples; the navigation along the longitudinal long flow path of the river Paisuni is limited only between Raghav Prayag ghat and Pramod Ban near Sitapur. The navigation at Raghav Prayag ghat will provide recreation for the tourists.

The Department of Town and Rural Planning, M.P. (1974) has proposed the navigation site between Raghav Prayag ghat and Jankikund. The site of navigation can be increased upto Sphatikshila by cleaning up the watershed of Paisuni river. It may be further increased upto Karwi town without any extra expenditure (Fig. 10.7). In the future plan, if the ravinous belt between Sitapur and Karwi have been environmentally checked and developed, the navigation proposal between Sphatikshila and Karwi Town will enhance the natural attraction and the concerned authorities will be highly benefited. The Chitrakut Development Plan, Town and Rural Planning Department, M.P. (1974) has also decided to maintain the beauty of ghats and to safe the bank of river Paisuni. It is also decided to provide plantations along the river course, beside the construction of ghats, temples and Dharamshalas along the river.

Ground Water Management

Ground water management is a significant issue in eco-management planning because it provides the socio-economic facilities and life support system of the region. The management plan incorporates the possibilities of ground water utilisation for existing as well as future generation. The ground water occurrences in different water bearing formations, the water level status (Fig. 1.20, e.g. ranges between 2.5 m and 57.47 m below ground level), the utilisation suitability and exploitation capacity (Fig. 1.21A) vary in different physiographic units of the region. The insufficient ground water potentials make the region as water scarcity zone. Obviously, it needs proper management through up-to-date conservation techniques. The pioneer field observations on the hydrogeological conditions as worked out by Bhatnagar, N.C. and Sah, D.L. (1967), Sinha, K.K. (1971), Srivastava, M.L. (1974), Sastry, V.P., Srivastava, V.N. and Singh, H. Saxena, V.N. and Garg, S.K. (1979) and Gupta, J.P. (1981) have also revealed that the ground water potentials of the region are very meagre.

On the basis of the Electrical Resistivity experiments carried out by the Geophysical Department, the ground water has been surveyed and utilized through wells, handpipes, tubewells, dug-cum-bore wells and dug-cum-blast wells for drinking as well as irrigational purposes. In spite of artificial exploitation of ground water, the springs and seepages are the natural exposures (Fig. 10.6) which provide the free supply of water facilities but deserve water management plans for efficient utilisation.

The following schemes may be suggested for ground water management of the region:

1. Following the Assessment Report for Rural Water supply scheme of Banda district, and District Statistical Report, the region faces hardship, because water is not available within 16 metres blow ground level and mostly dry up during the summer season. Handpumps and tubewells are almost absent in the plateau areas. Open wells are main source of drinking water available in about 91 per cent of the revenue villages. The District Statistical report reveals the fact that 5.81 per cent villages have not sufficient water during summer, 8.76 per cent villages have insufficient water throughout the year, 25.7 per cent of the villagers have to bring water from distant perennial streams and springs, 1.4 per cent bring water from ponds and tanks, 32.5 per cent from wells of distant places and 37.7 per cent from other sources such as railway water storage and the 'Bauli'. Thus, the region needs the management of sufficient water supply with the help of pipe systems connected with the artificial reservoirs of the area. The perennial source of ground water exploitation is the essential need in different water scarcity zones of the region which can be made available by the concerning departments.
2. The ground water quality assessment by State Hygiene Institute and Ground Water Directorate reveals the fact that the quality of water is mostly hard and brachish. It is found good in Raipura, Chitrakut, Bharatkup, Pahari, Karaundi and Khoh villages while moderate in quality near the villages like Karwi, Sheorampur and Anchwara respectively. In this connection, the maintenance of water quality standard, appointment of Chemist under Jal Nigam, collection of water samples and their chemical treatment, establishment of chemical laboratory in Karwi or Sitapur localities, appropriate use of bleaching powder in village well water etc. may help to maintain the water quality standard of the region.

3. The available data from district hospital, Banda denotes that the impact for water borne diseases like Typhoid (5% in Chitrakut, 40.7% in Manikpur, 0.15% in Ramnagar, 7% in Naraini and 0.7% in Pahari Buzurg), Amoebic and Bacillery dysentery (25.1% and 20.1% in Chitrakut Dham area), Hepatitis (16.7% in Pahari, Chitrakut Dham and Manikpur localities), paratyphoid, gasterites, enteritis and jaundice are generally marked in the region and adversely affected the health of the local people. To avoid pollution and to check the water borne diseases one should adopt some precautions in the construction of wells such as location of well site, grouting and scaling of well causing, completion of tap of well, disinfection of well, sanitary protection of pumping facilities, scaling the abandoned wells and moreover providing the facilities of primary health centre in the problem areas.
4. In view of provision of safe and adequate water supply of treated water, to minimise the rate of water borne diseases, to quench the thirst of scarcity villages in the summer, to improve the general health of rural masses and to stimulate the social and economic development it is necessary to plan the water supply scheme through pipe system as introduced earlier in 1973-74, the Patha rural water supply schemes for the Manikpur and Karwi areas. The connections should be made easily available for the rural poors.
5. The problems of short period of water supply, operation and maintenance problems, uncertainty of power supply, lack of public acceptance and participation due to high cost of connections bills and water tax and the harassment by the officials of the 'Jal Sansthan' require the proper remedial actions. The supply of water in tribal areas must be made available by low monthly charges.
6. The existence of spring and seepage conditions along the scarpline (Fig. 10.6) are the major natural exposures of ground water in the region, and the vital source of drinking water supply to the meagre population of the plateau region. People conserve water from such springs by constructing small rectangular tanks. It is suggested to construct large scale concrete tanks at the sites of the springs and water supply

must be managed by pipesystem in water scarcity zones as already arranged near Hanumandhara and Gupt-Godavari spring sites. The small tanks, constructed along the spring must be enlarged.

7. Although the state and central governments have made some surveys for assessing the potentials of ground water, the plateau area is yet to be benefited. Moreover, the equipments and techniques which are being used are old ones. In its place the new techniques like isolopes computers, geophysical equipments and Aerial photographs equipments and techniques will be used by Central Ground Water Survey Department and Ground Water Investigation Organisation. Hydrobotanic survey must be carried out to find out the depth of water table in different areas. In order to know the velocity and direction of ground water movements, Radiostopes with the help of carbon-14 and titanium can be best means in this regard (Gupta, J.P. 1981). Evaluation of a site for a well or boring pump should be done after meeting with local farmers, village heads and others so that a sense of involvement is created.

8. Pipe water supply scheme is a costly affair. Therefore, it is suggested that the scarcity villages should be covered by installing handpumps. Handpumps can be installed in rocky area by sophisticated drilling rigs as experienced in Karwi and Naraini areas.

9. The planning strategy must be made in social and commercial context. In order to save time and energy in relation to water collection journey and to improve rural health under the first stage benefits, the immediate aim should be the improvement of the quality, quantity availability and reliability of water (Feacham, R. 1975).

Tourism Development

Tourism development appears as an essential aspect of eco-management planning in a given region, because it helps to fascinate the morphological scenery of the existing terrain. The tourism processes involve to rejuvenate the natural eco-structure in an excellent and attractive form which in turn enhance recreational facilities and public interests with natural eco-system. Taking into consideration of prosperous natural properties and historico-spiritual contents of the region, activities of tourism development become necessary project under eco-management planning.

Chitrakut Dham is a famous pilgrimage centre due to its excellent natural beauty and remarkable hill sites of different rock exposure and rich vegetal covers. It is heartly to touch of Hindu mythology because 'Lord Rama' lived here during the period of exile. It has its great religious importance due to the birth place of Maharshi Balmiki and Goswami Tulsidas at Lalpur and Rajapur villages respectively and the famous 'Ashramas' like Maharshi Atri, Sati Anusuiya and Sutikshan along the Mandakini valley. It is also believed that Brahma, Vishnu and Mahesh were born in Sati Anusuiya Ashram. Historical evidences reveal the fact that Maulana Abdul Rahim Khan Khana, the greatest Hindi poet 'Keshavdas' and 'Pandavas' during the 'Agyatvas' have lived here for some times. All these religious affiliations attract the Hindu pilgrims of all the year around. The tourist centre and the holy 'Parikrama Marg' is closely associated with the hilly tracts of Sitapur and Karwi areas. Most of the 'Tirthas' are located either at the top of the plateau rim or along the Paisuni river where the topography is also peculiar and prominent in nature. Specially in Satna district, M.P. Kamadgiri Parvat is located both in U.P. and M.P.

The Chitrakut special area includes 18 villages having 143455 hectares of areal coverages (143.45 km^2) and 10187 rural population (1971). The temples of the region are not seen older than 200 years but historically it is older in age between 300 and 400 years (Chitrakut Development Plan). The development plan indicates that the average number of tourists per day exceed five thousand. A local fair on 'Amavashya' and 'Purnima' is held every months with an average attendance of pilgrims of 75000 and 20000 respectively. The special fairs are notable during the period of 'Dipawali', Ramanavami' and Somvati Amavashya parvas. About 1,50,000 pilgrims are estimated to come for Darshan of Lord Rama on the occasion respectively. Most of the tourists come from Banda, Hamirpur, Kanpur, Varanasi, Allahabad, Ayodhya, Manikpur, Satna, Rewa, Panna and Chhatarpur localities. Foreigners are not seen in the Parikrama marg. However, the Chitrakut region can be a good attraction for the foreigners too if all the cultural heritages are developed and decorated on scientific tourism lines. According to survey conducted by Town and Rural Area Planning Department, Chitrakut, it is apparent that the lodging and boarding facilities are unsatisfactory. The Dharamshalas are generally caste-oriented and the Rest Houses only provide the facilities for concerning officers. Traditional living facilities in 29 Dharamshalas are available in Satna localities while six Dharamshalas and one Tourist Bunglow are available in Sitapur, U.P. Thus the lodging and boarding facilities for tourists must be the essential part of development strategy.

The market facilities of the religion specifically in Dham area are in haphazard way. The tourists fulfil their needs in the market located along Mandakini valley near Raghav Prayag ghat. Therefore, a new advanced market centre is needed along the Kamadgiri Marg in Satna district. The local products (wooden toys and stony bowls) should be increased in the market centres.

Except a tourist office in Sitapur, there is not a single information centre from tourist point of view. The transportational facilities are rare. The tourist coach on taxies only provide the 'Darshan' of Sati Anusuiya, Jankikund, Pramodban, Sphatikshila, Gupt-Godavari and Sirsaban. The remaining 'tirthas' are located in the interior of the plateau where common tourist cannot be reached. The problems of water supply, sewage system, electric supply, security centres, post and telephone centres, petrol pumps and vehicle service centres are very common features must be removed in the area.

On the basis of 'Chitrakut Development Plan, Town and Rural Development Planning, Satna, M.P. the total number of tourists are estimated in 1991 as about 20 lakhs wherein 62.5 per cent (12.5 lakhs), 14.5 per cent, 9 per cent, 9 per cent (6.5 lakhs annually) and 5 per cent (daily) pilgrims are estimated during the 'Parvas' in Dham area. The 60 per cent tourists enter from U.P. sites while the remaining come from M.P. sites. It is estimated that 53 per cent, 28 per cent, 14 per cent and 5 per cent tourists stay for the period of one day, two days, three days and more than four days respectively. Taking into the consideration of tourists concentration, there are urgent need of 900 dwelling units in Dham area. About 15 buses are needed on Satna-Chitrakut marg while on the occasion of 'Parvas' it should be increased upto 50. The Planning Department has proposed to manage 8 tourist coach, 10 taxies and 15 local buses during the off season of the local fairs. A bus station for 33 vehicles is needed near Pramodban. The Planning Department has estimated about 3.8 hectares of excess land to be captured for vehicles and bus stand.

Being a famous religious centre, Chitrakut Dham area is facing a number of acute problems. Warning for new unauthorised constructions, improvement of road width in Parikrama marg, new plantation along the roads, public information centres, guide map and sites indicators on road sides, reconstructions of old bridges, temples, dharamsalas, roads and ghats, cleaning up the Paisuni course, making suitable the navigational sites, arrangement of advanced boarding and lodging facilities in Sitapur and Karwi areas, repairing of 560 stairs of Hanumandhara, making available the market and other infrastructural facilities etc. are the essential infrastructure which the Tourism department should take care of in a planned way. The Rajapur holy centre (the birth place of Goswami Tulsidas) should be improved. The proposal of 'Van Bihar' including parks and sanctuaries are acceptable.

CONCLUSION

On the basis of the overall discussion, it may be concluded that the Chitrakut region is environmentally degraded due to misleading activities of existing population and haphazard planning procedures of the state Governments. The socio-economic structure and infrastructural facilities may be minimised continuously if it is not managed carefully and protected scientifically.

The continuous degradation in topographical behaviour, the loss of soil fertility, the unwarranted ravination hazards, the clearance of vegetal properties, unchecked biological encroachment, mining and quarrying disasters, pollution impact in surface and subsurface water regimes as well as throughout the entire eco-system have already questioned the working efficiency of the officiating officials and planners of the region. For the sake of the existing eco-problems, the preceding proposals of planning may be creditable and fruitful. It is suggested to operate the plans honestly at governmental levels and to do beneficial efforts in a view of work and worship at local level. The socio-economic balance may not be maintained and improved until the ecological problems are not properly removed and redirected.

Bibliography

Ahmed, E., 1968: Distribution and Causes of Gully Erosion in India, Selected Papers, 21st International Geographical Union Congress, 1, pp. 1-3.

Ahnert, F., 1976: Brief Description of a Comprehensive Three Dimensional Process-response Model of Landform Development, *Journal of Geomorphology*, Supplementary Bd. 25, pp. 25-49.

Anderson, M.G. and Calver, A., 1977: On the Persistence of Landscape Features Formed by Large Flood, Transactions of Institute of British Geographers, N.S. 2, pp. 243-254.

Antev, E.V., 1952: Arroyo-cutting and Filling, *Journal of Geology* 60, New York, pp. 375-385.

Auden, J.B., 1933: Vindhyan Sedimentation in the Son Valley, Mirozapur District, *Mem. Geol. Surv. Ind.*, 62(2), pp. 141-256.

Balmiki Prasad, 1984: Geology, Sedimentation and Paleogeography of the Vindhyan Supergroup South-eastern Rajasthan, Pt. I, *G.S.I. Mem.* Vol. 116, pp. 67-102.

Banerjee, I.B., 1964: On the Broader Aspects of the Vindhyan Sedimentation, Report 22nd International Geological Congress, 15, pp. 189-204.

Banerjee, I.B. and Sengupta, S. 1963: The Vindhyan Basin: A Regional Reconnaissance of the Eastern Part, *Quart. Jour. Geol. Min. Metal Soc. Ind.*, 35, pp. 141-149.

Banerjee, I.B., 1974: Barrier Coastline Sedimentation Model and the Vindhyan Example, *Quart. Jour. Geol.* pp. 101-127.

Banerjee, I., 1982: *The Vindhyan Tidal Sea App. in Geology of Vindhyachal*, ed. by Validiya, Bhatia & Gaur, Hindustan Pub. Corp. New Delhi, pp. 80-87.

Bates, C.G. and Henry, A.J., 1920: Forest and Stream Flow Experiment at Wagon Wheel Gap Cola. Final Report on Completion of the Second Phase of the Environment, Monthly Weathering Review Suppl., 30, p. 79.

Bose, A., 1956: *Plant Life in the Vindhyan, Nature*, 178, pp. 927-928.

Bradley, W.C., 1963: Large-scale Exfoliation in Massive Sandstone of the Colorado Plateau, *Bulletin of Geological Society of America*, 74, pp. 519-528.

Brayan, K., 1941: Pre-Columbian Agriculture in South-west as Conditioned by Periods of Alluviation, *Annals of the Association of American Geographers*, 31, pp. 21-42.

Brunsden, D., 1979: *Mass Movement Appeared in Process in Geomorphology* (ed. by Embleton and Thornes, J.) Arnord-Heine Mann, Publishers (India) Pvt. Ltd., p. 130.

Cariston, C.W., 1963: Drainage Density and Stream Flow, U.S. Geol, *Survey Professional Paper*, 422C.

Carton, C.W and Langbein, W.B., 1960: Rapid Approximation of Drainage Density; Tine Interaction Method, *U.S. Geol. Surv. Wat. Res.* Division Bulletin, p. 11.

Chanda, S.K. and Bhattacharya, A., 1982: Vindhyan Sedimentation and Palaeogeography, Post Auden Developments, Appd. in Geology of Vindhyachal, ed. by Valdiya, Bhatia & Gaur, Hindustan Pub. Corp. New Delhi, pp. 88-101.

Chapman, C.A. and Rioux, R.L., 1958: Statistical Study of Topography, Sheeting and Jointing in Granite, Acadia National Park, Maine, *American Journal of Science*, 256, pp. 111-127.

Chorley, R.J., Malm, D.E.G. and Pogorzeiski, H.A., 1957: A New Standard for Estimating Basin Shape, *Amer. Journal of Science* 255, pp. 138-141.

Chorley, R.J. (ed.), 1969: *Water, Earth and Man*, Methuen, London.

Clowes, A. and Comfort, P., 1982: *Process and Landform. An Outline of Contemporary Geomorphology*, Oliver and Boyd, Robert Stevenson House, Edinburg, p. 16.

Cotton, C.A., 1964: The Control of Drainage Density; *N.Z. Journal, Geology and Geophysics*, Vol. 8, pp. 384-392.

Currecy, D.T., 1968: Bellfield Dam, Victoria Pt. 1 Site Geology, Inst. Engineers, Aust., Ann. Conf. Papers, pp. 133-136.

Davis, W.M., 1930: Origin of Limestone Caverns, *Bulletin of Geological Society of America*, 41, pp. 475-628.

Defant, A., 1958: *Ebb and Flow*, The University of Michign Press, Ann. Arbor., pp. 1-121.

Denevan, W.M., 1967: Livestock Number in 19th Century, New Maxico and Problem of Gullying in South-West, *Annals of the Association of American Geographers*, 57, pp. 691-703.

Dubey, R.S., 1968: *Erosion Surfaces on the Rewa Plateau, M.P., India*, Selected Papers, I.G.U. Calcutta, Vol. 1.

Dubey, V.S. and Chowdhury, M.S., 1952: Late Pre-Cambrian Glaciation in Central India, *Current Science*, Vol. 21, pp. 331-332.

Dwivedi, S.P., 1984: An Appraisal of Food Resources in Relation to Population in Banda Distt. Ph.D. Thesis (Unpublished) in Geog. Submitted in Kanpur University, Kanpur.

Eckholm, E., 1978: *Disappearing Species: The Social Challenge*, World Watch Paper 22.

Fairbank, E.E., 1925: A Modification of Lavaberg's Staining Methods, *Am. Mineralogists*, Vol. 10, pp. 126-127.

Faniran, A., 1971a: Implication of Deep Weathering on the Location of Natural Resources, *Nig. Geol. J.* 14(1), pp. 59-69.

Feacham, R., 1975: The Rational Allocation of Water Resources for the Domestic Needs of Rural Communities in Developing Countries, *Proc. of the Second World Congress on Water Resources*, Vol. 2, Health and Planning, New Delhi, p. 554.

Fermor, L.L., 1906: Note on the Occurrence of Gypsum in Vindhyan Series of Satna, *Records G.S.I.* Vol. XXXIII.

Fox, C.S., 1928: Contribution to the Geology of the Punjab Salt, Range Records 61, pp. 147-179.

Gardner, J., 1970a: Rockfall, A Geomorphic Process in High Mountain Terrain, *Albertan Geogr.*, 6, pp. 15-20.

Gardiner, V., 1971: A Drainage Density Map of Dartmoor, The Devonshire, ASSOC, Tr. Vol. 103, pp. 167-180.

Gardiner, V., 1974: A Photo Mechanical Technique for Production of Drainage Density Maps, *Cartogr. Jour.* Vol. 11, pp. 22-44.

Gardiner, V., 1979: Estimation of Drainage Density from Topographical Variables, *Water Resource Research*, No. 15, pp. 909-917.

Ghosh, D.B., 1976: The Vindhyan Basin in Bundelkhand Son Valley Region; In: Vindhyans of Central India, *Symp. Geol. Geol. Surv. India*, Bhopal (9th to 11th Nov. 1976).

Gilman, K. and Newson, M.D., 1980: *Soil Pipes and Pipeflow*, British Geomorphological Research, Group Research Monograph 1, Norwich, Geo Abstracts.

Gilvert, G.K., 1904: Domes and Domed Structures of the High Sierra, *Bulletin of Geological Society of America*, 15, pp. 29-36.

Gregory, K.J. and Park, C., 1974: Adjustment of River Channel Capacity Down Stream from a Reservoir, *Water Resource Research*, Vol. 10, pp. 870-873.

Griggs, D.T., 1936b: The Factor of Fatigue in Rock Exploitation, *Journal of Geology*, 44, pp. 781-796.

Hari Narain and Kaila, K.L., 1982: Inferences About Vindhyan Basin from Geophysical Data, Appeared in *Geol. of Vindhyachal*, ed. by Valdia, Bhatia & Gaur, Hindustan Pub. Corp., Delhi, pp. 179-192.

Heede, B.H. 1971: Characteristics and Processes of Soil Piping in Gullies, U.S. Deptt. of Agriculture and Forest Service Research Paper R.M. 68.

Henbest, L.G., 1931: The Use of Selective Stains in Palaeontology, *Jour. Palaeontology*, Vol. 5, pp. 355-364.

Henhel, J.S.A.W., Bayor and Coutts, J.R.H., 1938: Subsurface Erosion on a Natal Midlands, *Farm. S.A.J. Sci.* 35, pp. 236-243.

Heron, A.M., 1922: The Gwalior and Vindhyan System of Southeastern Rajputana, *Mem. G.S.I.* Vol. XIV (2).

Horton, R.E., 1932: Drainage Basin Characteristics, *Trans. American, Geophysical Union*, 13, pp. 350-361.

Horton, R.E., 1938: The Interpretation and Application of Run-off Plot Experiments with Reference to Soil Erosion Problems, *Proceeding of Soil Science Society*, 3, pp. 340-349.

Horton, R.E., 1945: Erosional Development of Streams and Their Drainage Basins; Hydrophysical Approach to quantitative Morphology, *Geol. Soc. Amer. Bull.* 56, pp. 275-370.

Hughes, P.J., 1972: Slope Aspect and Tunnel Erosion in the Loess of Banks Peninsula, Newzealand, *J. Hydrol.* (Nz) 11, pp. 94-98.

Hukku, B.M., 1963-64: Progress Report No. 1 and No. 2 on the Detailed Geological Investigations on the Anusuiya Sarovar Scheme, Banda Distt., U.P., G.S.I. (Unpublished).

Hutchinson, J.N., 1968: Mass Movement in Fairbridge R.W. *The Encyclopaedia of Geomorphology*, Reinhold New York, pp. 688-696.

Hutchinson, J.N., 1971a: *Mass Movement in Encyclopaedia of Earth Sciences* (ed. R.W. Fairbridge) Reinhold, New York, pp. 688-695.

James, H.L., 1954: Sedimentary Facies of Iron Formation, *Economic Geology*, Vol. 49, pp. 235-293.

Jones, J.A.A., 1971: Soil Piping and Stream Channel Infiltration, Water, *Resource Research*, 7, pp. 602-610.

Jayaswal, S.N.P. and Prasad, G., 1989: Types and Measurement Techniques of Morphological Processes, *AMSE Trans. AMSE Press France*, Vol. 4, No. 1, pp. 51-63.

Kedar Narain, 1959-60: Systematic Mapping in Parts of Karwi Tahsil, Banda District, U.P. G.S.I. Report (Unpublished).

Kedar Narayan, 1959-60: Geological Report of Karwi Area, G.S.I. Report (Unpublished).

Kedar Narain, 1960: Vindhyan Sedimentation in the Karwi Area, U.P. *G.S.I. Records*, Vol. 98, Pt. 2, pp. 96-106.

Keller, W.D., 1957: The Principles of Chemical Weathering, Lucas, Columbia, Miss.

Kiersch, G.A. and Asce, F., 1964: Vaiont Reservoir Disaster, *Civil Engineering*, 34, pp. 32-39.

King, L.C. 1962: *The Morphology of the Earth*, Oliver and Boyd, Edinburg, p. 699.

Kummar, N. and Sanders, J.E., 1976: Characteristics of Shoreface Storm Deposits, Modern and Ancient Examples, *Journal of Sedimentary Petrology*, 46, pp. 145-162.

Kumar, A., 1979: *Geomorphology of Simdega and Its Adjoining Area, Bihar*, N.G.S.I., B.H.U., Varanasi p. 50.

Lewin, J., 1970: *A Note on Stream Ordering, Area* 2, pp. 32-35.

Lewis, W.V., 1954: Pleasure Release and Glacial Erosion, *Journal of Glaciology*, 2, pp. 417-422.

Martin, P.S., 1963: *The Last 10,000 Years: A Fossil Pollen Record of the American South-West (TUCSON) Arizona*, University of Arizona Press, pp. 61-64.

Mccoy, R.M., 1970: Automatic Measurement of Drainage Networks, *IEEE Trans. Geol. Sci. Electron*, Vol. 8, pp. 178-183.

Mcclean, R., 1967: Origin and Development of Ridge-furrow System, in Beachrock in Barbados, West Indies, *Mar. Geol.*, 5, pp. 181-193.

Meddlicot, H.B., 1959: Vindhyan Rocks in Bundelkhand, *Geol. Mem.*, G.S.I., Vol. II.

Meddlicot, H.B. 1960: Vindhyan Rocks and Their Associates in Bundelkhand, *Geol. Mem.* 2, p. 37.

Mehrotra, C.L. and Gangwar, B.R.: *Soil Survey and Soil Work in U.P.*, Vol. 7, The Soil Survey Organisation, U.P. pp. 59-66.

Melton, M.A., 1957: An Analysis of the Relations Among Elements of Climate, Surface Properties and Geomorphology, Office of Naval Research, Technical Report II (Project NR389-042), p. 100.

Melton, M.A., 1958: Correlation Structure of Morphometric Properties of Drainage Systems and Their Controlling Agents, *Journal of Geology*, 66, pp. 442-460.

Miller, V.C., 1953: *A Quantitative Geomorphic Study of Drainage Basin Characteristics in the Clinch Mountain Area*; Va. and Tenn. Office of Naval Research Project NR 389-042, Tech. Rept. 3, Columbia University.

Mishra, R.C., 1949: Correlation of Bundelkhand Granites and Associated Rocks from Mahoba Area, Hamirpur Distt., U.P., *Proc. Ind. Sci. Congr. Assoc.* 36th Session 12.

Mishra, R.C., 1969: The Vindhyan System, *Proc. 56th International Science Congress*, 2, pp. 111-142.

Mosely, M.P., 1973: Rainsplash and the Convexity of Badland Divides, Z. *Geomorphological Suppl*. 18, pp. 10-25.

Muller, F., 1964: The Rock Slides in the Vaiont Valley, *Rock Mech. Engg. Geol*. 2, pp. 148-212.

Odum, E.P., 1971: *Fundamentals of Ecology*, W.B. Saunders Company.

Oldham, T., 1893: *Manual of Geology of India*, 2nd Edition, Manager of Printing, Govt. of India, Kolkata-5.

Ollier, C.D. and Tuddemham, W.G., 1962: Inselbergs of Central Australia, *Zeit. f. Geomorph*., 5, pp. 257-276.

Ollier, C., 1969: *Weathering*, English Language Book Society and Longman Group Limited London, p. 1.

Park, C.C., 1980: *Ecology and Environmental Management: A Geographical Perspective* (Butterworths), pp. 109-110; 195-196.

Pandey, R.S., 1983: A Geomorphological Study of Nagod and its Environ. Unpublished Ph.D. Thesis Submitted to Alld. University.

Parker, G.G., 1964: *Piping, A Geomorphic Agent in Landform Development of the Drylands*, Int. Assoc. Sci. Hydrology Assembly of Berkeley Publ. No. 65.

Peltier, L.C., 1950: The Geographical Cycle in Periglacial Regions As It is to Climatic Geomorphology, *Annals, Association of American Geographer*, 40, pp. 214-236.

Pichamuthu, C.S., 1971: Precambrian Geochronology of Peninsular India: *Jour. Geol. Soc. India*, 12(3), pp. 262-273.

Polynov, B.B., 1937: The Cycle of Weathering, Murby London, Trans. Alexander Muir, p. 220.

Potts, A.S., 1970: Forest Action in Rocks, Some Experimental Data, *Transaction of Institute of British Geographers*, 49, pp. 109-124.

Prasad, B., 1984: Geology, Sedimentation and Palaeogeography of the Vindhyan Supergroup, S.E. Rajasthan, *G.S.I. Mem*. Vol. 116, Pt. I, pp. 94-102.

Prasad, G., 1985: Environmental Crisis of Soil Erosion on Chitrakut Upland, *Proceeding, Int. Conf. Mod. Sim. Contl.* (Gorakhpur, India), AMSE Press France, Vol. 5, pp. 25-36.

Prasad, G., 1986: Morphometric Evaluation and Interpretation of Bifurcation Ratio in Paisuni Catchment Area, India, *Modelling, Simulation and Control*, AMSE Press France, C, Vol. 5, No. 2, pp. 1-12.

Prasad, G., 1986: Physical Properties, Chemical Analysis and Suitability of Groundwater in Vindhyan Formation of Chitrakut Plateau, *Mod. Sim. Contl.*, AMSE Press France, C, Vol. 5, No. 2, pp. 1-12.

Prasad, G., 1987: Low and Order of Hydrological Concept Regarding the Drainage Density and Drainage Density Ratio Under the Effective Environmental Controls of Paisuni-Gauta Interfluvial Zones; *Moddling, Sim. Control.* AMSE Press France C, Vol. 7, No. 2, pp. 14-32.

Prasad, G., 1987: Law of Stream Number of Selected Drainage Models of Chitrakut Upland, India, *Modelling, Simulation and Control*, AMSE Press France, C, Vol. 7, No. 4, pp. 25-31.

Prasad, G., 1987: Geometry of Shape of Closed Links: A Morphometric Approach for 20 Selected Model of Drainage Basins of Chitrakut Upland, *Mod. Sim. Contl.*, AMSE Press France, C. Vol. 8, No. 4, pp. 49-59.

Prasad, G., 1987: Modelling of Morphogenetic Processes of Chitrakut Upland, *AMSE Transactions*, AMSE Press France, Vol. 1, No. 2, pp. 1-12.

Prasad, G., 1987: Measurement of Complexity of Relationship Between Average Slope and Associated Morphometric Variables of Chitrakut Upland, India, *Mod. Sim. Contl.* AMSE Press France, C. Vol. 8, No. 2, pp. 35-46.

Prasad, G., 1987: Modelling of Major Drainage System and Environmental Control of Chitrakut Upland, *Mod. Sim. Contl.* AMSE Press France C. Vol. 8, No. 1, pp. 52-62.

Prasad, G., 1987: Modelling of Topographic Variations and Physiographic Regionalisation of Chitrakut Upland, *Mod. Sim. Contl.*, AMSE Press France C, Vol. 8, No. 3, pp. 1-9.

Prasad, G., 1988: A Comparative Study of Relative Relief of Chitrakut Upland, *AMSE Trans*. AMSE Press France, Vol. 2, No. 3, pp. 1-26.

Prasad, G., 1988: Measurement of Drainage Dissection and Environmental Control, *AMSE Transactions*, AMSE Press France, Vol. 2, No. 3, pp. 1-26.

Rabinson, H., 1972: Biogeography, Macdonald and Evans London.

Rathjens, C., 1973: Substerrange Abtragung (Piping) *Z. Geomorph. Suppl*. 17, pp. 168-176.

Ravi Prakash and Delela, I.K. 1982: Stratigraphy of the Vindhyan in Uttar Pradesh: A Brief Review, Appd. in *Geology of Vindhyachal*, ed. by Valdiya, Bhatia & Gaur, Hindustan Pub. Corp. New Delhi, pp. 55-79.

Reiche, p., 1950: *A Survey of Weathering Processes and Products*, New Maxico, University Publication Geology-3.

Reinson, G.E., 1976: Facies, Models, Barrier Island System, In Walker, R.G. (ed.) *Facies Models*, Geoscience Canada Reprint Series, 1, pp. 57-74.

Rode, K.P., 1946: A New Kind of Fossil in Vindhyan Rocks of Rohtas Hills in Bihar, *Curr. Sci*. 15 (9), pp. 247-248.

Roy, A.K. and Bhattacharya, A., 1982: Regional Geomorphology of Vindhyachal, Appd. in *Geology of Vindhyachal*, ed. by Valdiya, Bhatia & Gaur, Hindustan Pub. Corp. New Delhi, pp. 9-22.

Rubin, J., 1966: Theory of Rainfall Uptake by Soils Initially Drier Than Their Field Capacity and Its Application, *Water Resource Research*, 2, pp. 739-794.

Safaya, H.L., 1963-66: On the Vindhyan Dolomite of Banda Distt., *G.S.I. Report*, Vol. 106, Pt. 2, pp. 110-124.

Salujha, S.K., 1982: Palynology of the Surface and Sub-surface Vindhyan Sediments, Appeared in *Geol. of Vindhyachal*, ed. by Valdia, Bhatia & Gaur, Hindustan Pub. Corp., Delhi, pp. 113-122.

Scheonwetter, J., 1962: The Pollen Analysis of Eighteen Archaeological Sites in Arizona and New Mexico, Chapter 8 in the Chapters in the *Pre-History of Eastern Arizona Fieldiana Anthropology*, Vol. 53.

Schidegger, A.E., 1965: The Algebra of Stream-order Numbers, *U.S. Geol. Survey*, Prof. Paper 525B, B1, pp. 87-89.

Schidegger, A.E., 1966: Stochastic Branching Processes and the Law of Stream Orders, *Water Resources Research* 2, pp. 199-203.

Schumm, S.A., 1956: The Evolution of Drainage System and Slopes in Badlands at Perth Amboy, New Jersey, *Geol. Soc. Amer. Amer. Bull.*, 67, pp. 597-646.

Schumm, S.A., 1963: A Tentative Classification of River Channels, *U.S. Geological Survey Circular* 477, p. 10.

Sharma, R.P., 1982: Lithostratigraphy, Structure and Petrology of the Bundelkhand Group, Appd. in *Goel. of Vindhyachal*, ed. by Valdia, Bhatia and Gaur, Hindustan Pub. Corp. Delhi, pp. 30-46.

Shreve, R.L. 1966: Statistical Law of Stream Numbers, *Journal of Geology*, 74, pp. 17-37.

Shreve, R.L., 1967: Infinite Topologically Random Channel Networks, *Journal of Geology*, 75, pp. 178-186.

Shrock, R.R., 1948: *Sequence in Layered Rocks*, McGraw Hill New York.

Siffer, V.V., 1977: *Physical Geology*, Mir Publishers Moscow, p. 125.

Singh, I.B., 1980: The Bijaygarh Shale Vindhyan Stream (Precambrian) India, An Example of Lagoonal Deposit, *Sedimentary Geology*, 25, pp. 83-903.

Singh, S., 1976: On the Quantitative Parameters for the Computation of Drainage Density, Texture and Frequency: A Case Study of a Part of the Ranchi Plateau, *National Geographer*, Vol. 10, Issue 1, pp. 21-31.

Singh, S.N. and Pal, O.P., 1970: Geology Around Chitrakut Area, Distt. Banda, U.P., *Journal of the Paleontological Society of India*, Vol. 14, pp. 77-85.

Sitheley, R.V., Srivastava, P.N. and Verma, C.P., 1953: *Microfossils from the Upper Vindhyan Area: A Discussion on the Age of the Vindhyan in Light of Plant Fossils*, Proc. Nat. Ind. Sci. Ind. 195-202.

Smart, J.S., 1967: A Comment on Horton's Law of Stream Numbers, *Water Resources, Research*, 3, pp. 773-776.

Woldenberg, M.J., 1966: Horton's Law Justified in Terms of Allometric Growth and Steady State in Open Systems, *Geol. Amer. Bul*l., 77, pp. 431-434.

Young, A., 1963: Deductive Models of Slope Evolution, Slopes Commercial Report, 3, pp. 45-66.

Index